This book is to be returned on or before
the last date stamped below.

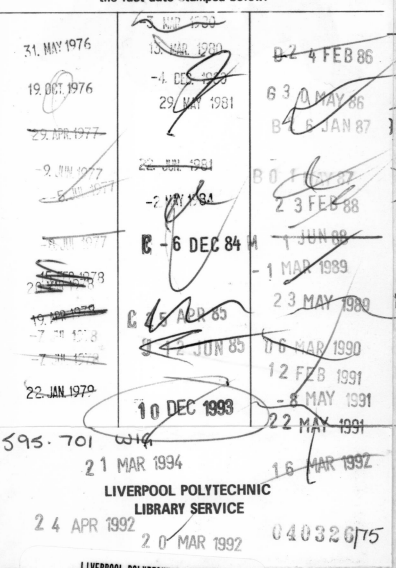

INSECT PHYSIOLOGY

Insect Physiology

V. B. WIGGLESWORTH C.B.E., M.D., F.R.S.

Emeritus Professor of Biology,
University of Cambridge.
Formerly Director, Agricultural Research
Council Unit of Insect Physiology

SEVENTH EDITION

CHAPMAN AND HALL London

First published 1934
by Methuen & Co Ltd
Second edition 1938
Third edition 1946
Fourth edition 1950
Fifth edition 1956
Sixth edition 1966
Reprinted 1967
Seventh edition published 1974
by Chapman and Hall Ltd
11 New Fetter Lane London EC4P 4EE

First published as a Science Paperback 1966
Reprinted 1967
Second edition 1974

Printed in Great Britain by
Willmer Brothers Limited, Birkenhead

SBN 412 11150 0 casebound
SBN 412 20980 2 paperback

Distributed in the U.S.A.
by Halsted Press, a Division
of John Wiley and Sons Inc., New York

Library of Congress Catalog Card Number 74–1548

Contents

haemolymph. Accessory functions of the Malpighian tubules.

To the insect, order and disorder are exposed to sight
and so we think to see the little emmets confer
and locking their antennae immediatly transmit
the instinctiv calls which each and all can feel; whereas
the mutual fellowship of distributed cells
hath so confronted thought that explanation is fetch'd
from chemic agency; because in that science
the reaction of unknown forces is described and summ'd
in mathematic formulae pregnant of truth,
and of such universal scope that, being call'd laws,
their mere description passeth for Efficient Cause.

—ROBERT BRIDGES

Introduction

The fundamental processes of vital activity, the ordered series of physical and chemical changes which liberate energy and maintain the 'immanent movement' of life, are probably the same wherever 'living matter' exists. The description of these changes is the ultimate goal of physiology, of whatever group of organisms; but in this book we shall be concerned only with physiology on a humbler plane: with the grosser functions of the organs and tissues, and with the mechanisms by which these functions are co-ordinated to serve the purpose of the insect as a whole.

Of all the zoological classes, the insects are the most numerous in species and the most varied in structure; the general physiology of the group is therefore only too apt to be obscured by the endless specializations of particular forms; and the main difficulty of a work like the present is the exclusion of all that is special and non-essential, and the retention of only that material which best illustrates the general theme. There are, indeed, certain common factors which condition the physiological make-up of the insects, and these factors serve, to some extent, to link them all together into one system. They are essentially terrestrial animals. This circumstance determines the characters of their cuticle, and this, in turn, conditions their respiratory mechanism and the physiology of their growth. The respiratory mechanism and the cuticular skeleton are

vii

among the factors which restrict the size of insects. Their small size and terrestrial habit render them very prone to lose water; and the urgent need for the conservation of water influences the respiratory, excretory and digestive systems. All these systems show special changes when the insect reverts to an aquatic life.

Yet, although the interaction of these factors results in some uniformity of principle in the physiology of insects, their diversity in habit, food and environment causes such endless variation in detail, that, quite apart from the conspicuous gaps in our present knowledge, any limited treatment of the subject must in any case be more or less arbitrary in presentation; and, were there space to insert them, nearly all the generalizations that are attempted should have qualifying instances.

The history of insect physiology is peculiar. The early microscopists, Hooke, Malpighi, Leuwenhoek, made many observations on the structure of insects, and many accurate inferences about their physiology. More was added by the great naturalists, Swammerdam, Réaumur, de Geer. But, with such conspicuous exceptions as Newport, Graber, Lubbock, Plateau, and others of more modern times, the majority of entomologists, until recent years, have been so fully occupied with the morphology and taxonomy of their colossal group that such advances in physiology as have been made have commonly been mere by-products of morphological study. From time to time we find the physiologists of the last century, Dutrochet, Treviranus, Marshall Hall, von Kölliker, Claude Bernard, turning to the insects to illuminate their theme; but their concern was not with the insect as such.

Within recent years, interest in the physiology of insects has arisen in a new quarter. The applied entomologist, confronted with the ravages of insects in the spheres of agriculture and of public health, has wanted to know something about their nutrition, about the laws governing their responses to sensory stimuli, about their reactions to parasites, about the precise way in which their bodies are adapted to diverse climatic conditions, and about the action upon them of toxic sprays and gases. With this demand for increased knowledge has come a realization of our present ignorance.

Since the first edition of this book was published almost forty years ago extensive advances have been made in all parts of the subject. Full-sized textbooks have been devoted to it; notably *The Principles of Insect Physiology* by the present author, the three volume *Physiology of the Insecta* edited by Morris Rockstein, and *Insect Biochemistry* by Darcy Gilmour; and articles describing the most recent advances in the physiology and biochemistry of insects appear in the *Annual Review of Entomology*, in *Advances in Insect Physiology* and elsewhere. References in this edition have therefore been confined to such textbooks and reviews, to a few recent papers which have not yet become incorporated in this way, and to a limited number of other papers which provide useful starting-points for further reading.

1 The Integument

The key to much of the physiology of insects is to be found in the nature of their cuticle. As was first shown by Haeckel, the cuticle is the product of a single layer of epidermal cells. It is often described as being composed of non-living material; but in fact the epidermal cells give off fine filaments contained within the so-called 'pore-canals', which run through the substance of the cuticle and often come within less than a micron of the surface.

Cuticle Structure

As described from stained sections the cuticle consists of two primary layers, the *endocuticle* which makes up the greater part, and a thin refractile *epicuticle* on the surface, usually not more than one micron in thickness. In the harder regions of the integument the outer half or third of the endocuticle is converted to a deep brown or amber layer, termed the *exocuticle*. In stained sections the epicuticle and exocuticle appear homogeneous and structureless; the endo-cuticle shows more or less conspicuous horizontal lamellae. The pore canals can rarely be seen. But if the cuticle of certain insects, such as *Periplaneta, Tenebrio, Rhodnius,* &c., is cut in the fresh state with the freezing microtome and mounted in water without staining, the pore-canals are conspicuous objects (often far more conspicuous than the horizontal lamellae) running right through the endo- and exocuticle. Sometimes their contents may be hardened in the exo-cuticle of the mature insect; but very often, if the fresh sections are dried in air before being mounted, their moist contents contract and

1

the canals become filled with fine threads of gas. Clearly the cuticle is far from being dead: it often contains threads of cytoplasm coming almost to the surface; in this respect it resembles the bone and dentine of vertebrates.

The 'living' character of the cuticle is evidenced by the fact that under the action of some humoral factor conveyed by the nerves it will become more pliable and extensible after feeding; and a hormone from the central nervous system may control its hardening and darkening at the time of moulting.

Composition of the Cuticle

Chitin. The most familiar constituent of the cuticle is the nitrogenous polysaccharide chitin; but this rarely forms more than 50 per cent of the substance of the cuticle. It is most abundant in the flexible and elastic endocuticle; it is not responsible for the hardness of the exocuticle, where it is reduced in amount, and it is entirely absent from the epicuticle. Chitin, a polymer of acetylglucosamine, is closely related to cellulose, and like cellulose it exists in the form of sub-microscopic crystallites or micellae. In the endocuticle these tiny rodlets are aligned to form fibrils and these fibrils tend to be oriented all in one direction at one level in the cuticle. They may retain this 'preferred' direction in one lamina of the cuticle and then take up a new preferred direction in the succeeding lamina, at an angle of 60° or so to the first. In other cuticles the orientation of the fibrils changes systematically in each successive layer to form a regular spiral. The fibrils of chitin are bound together by a protein matrix after the manner of fibre glass; indeed, the two may well be chemically united to form a mucopolysaccharide.

Protein. The protein of the endocuticle is readily demonstrated by the protein colour tests: the biuret, Millon's and xanthoproteic reactions, all of which are strongly positive. The best test for chitin is that of van Wisselingh, which consists in its conversion into chitosan by saturated caustic potash at a high temperature, and the recognition of chitosan by its solubility in acids and by the violet colour it gives with iodine.

Sclerotin. In the exocuticle the protein component has become converted to a horny substance termed 'sclerotin'. Sclerotin bears some resemblance to vertebrate horn or keratin; but whereas keratin is described as 'vulcanized protein', a substance in which adjacent protein chains are chemically bound together by means of sulphur linkages, sclerotin is described as 'tanned protein'. It is produced by the action of quinones which are formed in the cuticle from the oxidation of various diphenols. The quinones react with the free amino groups in adjacent protein chains and bind these firmly together, converting a soft, white, extensible material into a hard and horny substance that may vary in colour from amber to deep brown. Gelatin tanned with benzoquinone gives rise to somewhat similar material. It was the invention of sclerotin and keratin which made possible a truly terrestrial existence for insects and vertebrates. It is these same materials which have made possible the development of wings. It is probable that other milder forms of polymerization, some involving the inclusion of lipids, occur in the cuticular proteins.

Resilin. The ordinary protein of the cuticle, which is tanned to form sclerotin, is called 'arthropodin'. In addition there is always present another protein called 'resilin'. This is an elastic substance in which the protein chains are bound together in a uniform three-dimensional network so as to form a perfect rubber. Resilin may be deposited between the chitinous lamellae and provides the elasticity of the cuticle; or it may exist in the pure state to form elastic hinges at the basis of the wings, and elastic tendons for muscles connected to the wings or elsewhere.

Lime. In a few aquatic larvae of Diptera there may be deposits of lime in the cuticle and in a few Diptera the puparium may be strengthened with a deposit of lime inside. But insects in general do not harden the cuticle as do many Crustacea by incorporating lime in the form of calcite. It is interesting to note, however, that sclerotin is actually harder than calcite, and that for their mandibles and claws which they need as hard as possible the Crustacea provide a covering of sclerotin. The sclerotized mandibles of some insects are so hard that they can readily bite through sheets of foil of lead, copper, tin, zinc or silver.

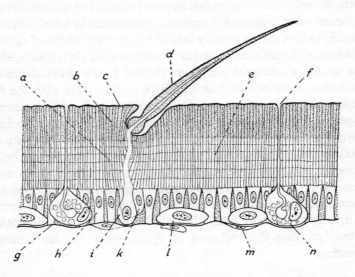

Figure 1. Section of typical insect cuticle; the sense cell and nerve fibril supplying the tactile bristle are omitted

a, laminated endocuticle; *b*, exocuticle; *c*, epicuticle; *d*, bristle; *e*, pore-canals; *f*, duct of dermal gland; *g*, basement membrane; *h*, epidermal cell; *i*, trichogen cell; *k*, tormogen cell; *l*, oenocyte; *m*, haemocyte adherent to basement membrane; *n*, dermal gland

The Epicuticle

The endocuticle is responsible for the extensibility of the integument and for combining toughness with flexibility. The exocuticle provides the rigidity in the hard parts such as the head capsule, the segments of the limbs, and so forth. The epicuticle is responsible for the impermeability of the cuticle, and particularly its power of preventing the loss of water by evaporation.

When the epicuticle of the fully developed integument is examined microscopically it appears as a refractile amber-coloured layer, usually not more than a micron in thickness. It is inextensible; but where the cuticle is liable to bend or stretch it is deeply folded.

The epicuticle is a highly complex structure. Its nature certainly varies in different insects, and the only way in which it has been possible to learn something of its composition has been to observe the stages in its development when a new cuticle is being formed.

Cuticulin. At this time the epidermal cell first lays down an 'outer epicuticle', a uniform layer of lipoprotein about 18 nm thick, and below this a much thicker 'inner epicuticle' of $0·5-1$ μm, also containing protein and lipid, before the laminated chitin-containing cuticle is formed. The polymerized lipoprotein which composes the epicuticle has been named 'cuticulin' and similar material may also impregnate the exocuticle. Indeed, 'cuticulin' may well be incorporated into the tanned 'sclerotin' of the cuticle. When it is first formed the epicuticle is pierced by the pore-canals and from these there exudes upon the surface a viscid secretion. Shortly before moulting a crystalline layer of wax separates out over the surface of this secretion and serves to waterproof the cuticle. And finally, at the time of moulting or soon afterwards, little glands (dermal glands) pour out a protective covering of secretion, called the 'cement layer', which spreads evenly over the wax. The nature of the cement layer is not fully known, but there is evidence that in the cockroach and other insects it contains resinous material resembling shellac.

Wax. The waxes of the epicuticle play a most important part in protecting the insect from loss of water. They vary in character from soft greasy materials to hard crystalline substances like beeswax. If the insect is exposed to a high temperature (35° C. in the cockroach, 55° C. in the (dead) *Rhodnius*) the crystalline layer of wax, which is of submicroscopic thickness, begins to loosen, and the rate of water loss increases rapidly. The wax is secreted by way of the pore canals and the insect is rendered waterproof before moulting. The pore canals end below the epicuticle; but they contain bundles of exceedingly fine filaments, the 'wax canal filaments', about 8nm thick and visible only with the electron microscope. These tubular filaments probably transmit wax or wax precursors from the cells to the surface of the cuticle. At the termination of the pore canal they spread out fanwise and pass right through the epicuticle. Other waxes may be secreted in association with protein and then gradually crystallize out – but the precise mechanisms are not yet fully understood.

Surface Forces and the Insect Cuticle

Since the surface of a body varies with the square of its radius, whereas its volume or mass varies with the cube of its radius, it follows that the ratio of surface to mass become progressively greater as the size of the body diminishes. Consequently, the forces resident in the surface become relatively stronger. In the case of insects, the surface forces are, indeed, frequently great enough to move the whole mass of the body; and they introduce a factor into the life of insects, particularly when they come into contact with water, which is unnoticed by larger animals. It is necessary, therefore, to consider the properties of the cuticle from this point of view.

Wetting properties. As a general rule, the cuticle is not readily wetted by water; there is, in fact, a large 'angle of contact' between the water surface and the cuticle (Fig. 2, A). Consequently, when an insect stands on water, the surface tension acts in the opposite direction to gravity, and the insect is held up as though by an elastic membrane. This 'hydrophobe' property may belong to the epicuticle; but in some cases at least it is accentuated by the presence of glands producing a fatty secretion. That is the case, for example, around the spiracles of mosquito larvae, and around the respiratory tube of the larva of *Eristalis*, by means of which this insect, like many other aquatic larvae, is enabled to hang suspended from the surface film with the remainder of the body submerged (Fig. 2, B).

Hydrofuge hairs. The effect of surface forces becomes still more evident when hairs or bristles are present on the body. When a fine pile of these hydrophobe hairs is set vertically on the body surface, they hold the water at a distance, so that the insect, when it goes below the water surface, can carry with it a film of air (Fig. 2, E, F). We shall see that this film has some remarkable properties in relation to the respiration of aquatic forms (p. 28). If a single bristle is inclined at an angle – for simplicity we will suppose an angle of 45°, and we will suppose that the angle of contact is 135° – and brought up to the water surface from below, as can be seen from Fig. 2, C, the surface tension of the water will draw it over so as to increase the angle of inclination. It will behave as though the one surface were hydrophile and the other hydrophobe. Many insects hold themselves at the water surface for purposes of respiration by means of rows or circles of such 'semihydrofuge' bristles (Fig. 2, D).

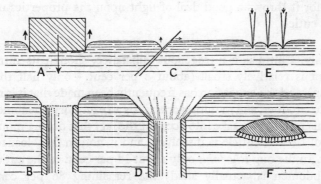

Figure 2. Diagrams illustrating the properties of the insect cuticle in contact with water

A, hydrophobe object supported by the surface tension of water; B, respiratory siphon with hydrophobe rim similarly supported; C, surface forces acting on a hydrophobe hair brought obliquely to the water surface; D, a crown of such hairs supporting a respiratory siphon; E, hydrophobe hairs brought vertically against the water surface; F, insect body with a ventral covering of such hairs carrying a layer of air

Adhesive organs. Terrestrial insects avail themselves of surface forces in quite a different way: namely, in order to cling to surfaces too smooth to provide a purchase for their claws. The physical mechanism of the organs (pulvilli and such-like) that are used for this purpose is by no means certain in all cases; but in one example that has been studied in detail, that of the bug *Rhodnius*, it has been shown that the insect is held by the adhesion or seizure of the climbing organ to the surface. In other words, it is the surface molecular forces that are directly concerned in supporting the body weight.

The Moulting, Formation and Pigmentation of the Cuticle

Through the possession of a more or less rigid carapace, incapable of increasing in surface area once it has been laid down, the insect incurs the necessity for discontinuous growth. In later pages (p. 96) we shall consider some of the physiological processes which lie at the back of insect growth; but one change which it always involves is the casting of the old cuticle and the formation of a new and larger one. This side of growth, or ecdysis, may be conveniently considered here, for it throws a good deal of light upon the properties of the insect cuticle.

Ecdysis. Chitin is a nitrogenous substance, though its nitrogen content is relatively small (about 8 per cent. – only half that of protein), and the suggesion has frequently been made that it is really a waste product, and moulting or ecdysis an excretory phenomenon. But this idea fails to take into account the fact that during moulting most of the old cuticle is dissolved, and is not, in fact, lost to the insect. In the bug *Rhodnius* only about 14 per cent. of the cuticle of the abdomen is actually shed; the remaining 86 per cent. is reabsorbed by the insect and is presumably used again for the formation of the new cuticle.

Moulting fluid. During this process only the endocuticle is dissolved, the chitinous and protein layer; the exocuticle and epicuticle are untouched. Solution is brought about by means of the 'moulting fluid'. This is a thin layer of liquid which lies between the newly developing cuticle and the old cuticle that will be cast off. This liquid contains enzymes, a chitinase and a proteinase, which will digest the substance of the endocuticle but will not touch the cuticulin of the epicuticle or the sclerotin of the exocuticle. Thus the new cuticle is protected from its action by the epicuticular covering. The enzymes of the moulting fluid appear to be secreted from the general surface of the new cuticle; perhaps they are discharged at the tips of the pore canals. And the products of digestion are likewise absorbed through the general surface which at this stage is still quite permeable. And all the time this process is going on the epidermal cells are synthesizing and depositing the chitin and protein of the new endocuticle; indeed, the deposition of these inner layers may continue for several weeks after the old cuticle has been shed.

Cuticle hardening. Up to the time of moulting the new cuticle is soft and almost colourless. Within an hour or two of the casting of the old skin it has both hardened and darkened. The mechanism of hardening, the quinone tanning of the cuticular protein to form sclerotin, and other forms of polymerization, have already been described. This process involves a considerable degree of darkening, so that the two processes go forward together. In addition, there are probably other oxidative processes going forward in the blackest parts of the cuticle leading to the production of melanin from the amino-acid tyrosine.

We are now in a position to appreciate the activities of the epidermal cell during the formation of the new cuticle. This cell secretes the cuticulin of the epicuticle, the chitin, the 'arthropodin' and 'resilin' and other proteins of the inner layers, the chitinase and protease which dissolve the old skin, then the phenolic material which will give rise to the quinones, and the waterproofing wax. Finally, it produces the oxidase which completes the hardening process. And all these activities are exquisitely timed (and synchronized throughout the body) to follow one another within a matter of hours or minutes. Indeed, in the formation and maintenance of its little share of cuticle the epidermal cell has shown itself a chemical agent with an amazing range.

The Mechanism of Moulting and of Hatching

Ecdysial lines. When the insect has digested its old cuticle and laid down the new one, it has fresh problems to solve. In the first place, it has to rupture the old skin and escape from it. Now any cuticle that has to be shed, that is, the cuticle in every stage except the adult, has an ecdysial line, or 'line of weakness' – sometimes in the head, most often in the midline of the thorax. Along this line the exocuticle is wanting, and the endocuticle extends right up to the very delicate epicuticle (Fig. 3, A). Consequently, when the endocuticle is completely dissolved (and the process of moulting is so timed that this happens just when the new cuticle is ready) the old skin needs but the slightest force for it to split along the line of weakness (Fig. 3, B).

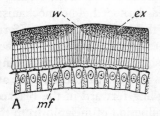 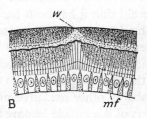

Figure 3. Transverse sections through the prothorax of an insect.

A, the new epicuticle is formed; digestion of the old endocuticle has scarcely begun.

B, digestion and absorption of the old endocuticle almost complete; along the ecdysial line only the epicuticle remains.

ex, exocuticle; *mf,* moulting fluid; *w,* ecdysial line of weakness

Mechanism of ecdysis. The rupturing force is the pressure of the blood. The insect contracts the abdominal muscles, often repeatedly in a peristaltic fashion, and so increases the pressure of blood in the thorax, which bulges outwards until the old skin splits. Many insects add to the efficiency of this process by first swallowing air (as vertebrates fill the lungs with air during defaecation or parturition) or, if they are aquatic, by swallowing water. Once the old skin is split, the insect escapes from it by gentle peristaltic movements, often aided by gravity – for many insects hang head downwards during moulting.

It is commonly stated that the moulting fluid serves as a lubricant to facilitate this process – indeed, that is often considered to be its chief function. But in many insects the moulting fluid has practically disappeared by this stage; the insect is quite dry, the inner surface of the old skin only slightly moist; and in many aquatic and terrestrial pupae a layer of air appears before eclosion, between the pupal skin and the developed adult – due, no doubt, to the energetic absorption of the moulting fluid.

Expansion of new cuticle. When the insect has withdrawn its limbs, and freed itself completely from the old skin or from the pupa, it must enlarge the new cuticle rapidly to the required size, and bring about the expansion of its wings, before the epicuticle has hardened. This it does in the same manner, by swallowing more air or water, and forcing the body fluid under pressure into the wings. In the Ephemeroptera and in many Lepidoptera the sole function of the gut in the adult insect is to receive air for this purpose. If the crop of the cockroach is pricked with a needle, the distension of the new cuticle cannot take place, and the insect collapses like a pricked balloon; or if the tips of the wings in the dragon-fly are cut off, blood drips from them and they cannot expand.

Hatching mechanism in the egg. We may note here that the process of hatching from the egg is similar in many particulars to that of moulting or emergence from the pupa. In the mature embryo of grasshoppers there are glandular organs ('pleuropodia') on the first abdominal segment which secrete enzymes that dissolve the inner layers of the shell. The amniotic fluid in which the embryo is bathed may be in part absorbed through the cuticle of the young insect; but in many cases the fluid is swallowed by the mouth shortly before hatching (like the moulting fluid of the silkworm). Many insects then tear or bite their way out of the egg. Others swallow air, and by muscular contraction of the abdomen, drive the head or thorax against the shell and cause it to burst open. Certain groups have special structures, spines, saws, hard plates or distensible bladders, which enable them to increase the efficiency of this process by concentrating the impact at one point of the egg, and so piercing the shell or forcing off the egg cap (Fig. 4). (This last arrangement finds a parallel in the ptilinum of emergent Muscid flies, which they distend with blood in order to force off the cap of the 'puparium'.) When the insect has left the egg, it commonly swallows air or water once more, and so increases its bulk before its cuticle has hardened.

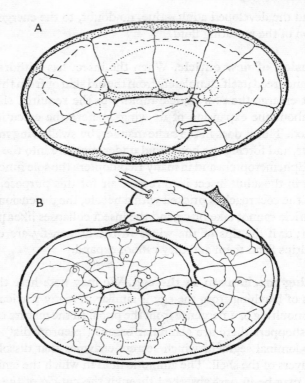

Figure 4. Hatching of the egg of the flea (after Sikes and Wigglesworth). In A the fully developed larva is still bathed in the amniotic fluid which it is in process of swallowing. Tracheal system still filled with fluid. In B, the larva has moved round within the egg and has cut a slit in the chorion with its hatching spine. All the amniotic fluid has gone and the tracheal system has filled with air

Eclosion hormone and 'bursicon'. The eclosion of some insects, such as the emerging adult of the silkmoth *Antheraea*, is induced by an 'eclosion hormone' liberated by neurosecretory cells in the brain, which brings into action all the necessary changes in behaviour. The complex series of enzyme activities in the cuticle that lead to hardening and darkening after eclosion are likewise set going by a hormone (sometimes named 'bursicon' because of its connection with tanning) which is produced by neurosecretory cells in the brain or elsewhere in the central nervous system and acts upon the epidermal cells.

Metallic Colours of the Cuticle

In many insects the surface of the cuticle itself, or of the flattened scales which are articulated to it, show brilliant metallic colours. In most cases these colours result from some periodic structure in the cuticle. In at least one insect, the beetle *Serica sericea*, the faint iridescence is due to diffraction by fine striae: collodion impressions of the wing-case, bearing a cast of this structure, show the same iridescence as the wing-case itself, But in the majority of insects iridescence is the result of interference by multiple thin films separated by material of slightly different refractive index. Horizontal thin plates of this kind are responsible for the metallic colours of Lycaenids and many other butterflies. The brilliant blue of *Morpho* butterflies is produced by thin plates which are inclined at an angle to make up the glassy ridges that run lengthways along the scales. These periodic structures presumably arise spontaneously by crystallization within the structure of the cuticle: ordinary Lepidopterous scales show periodic projections apparently homologous with those which produce the colours in *Morpho*.

BIBLIOGRAPHY

ANDERSEN, S. O. *J. Insect Physiol.*, **18**, (1972), 527–540 (hardening of locust cuticle)

ANDERSEN, S. O. and WEIS-FOGH, T. *Adv. Ins. Physiol.*, **2**, (1964), 1–66 (the rubber-like protein 'resilin' of insect cuticle)

BEAMENT, J. W. L. *Biol. Rev.*, **36**, (1961), 281–320 (water relations in insect cuticle: review)

COTTRELL, C. B. *Adv. Ins. Physiol.*, **2**, (1965), 175–218 (hardening of insect cuticle: review)

HACKMAN, R. H. *Physiology of Insecta*, III. (Morris Rockstein, Ed.) Academic Press New York, 1964, 471–506 (chemistry of insect cuticle)

LOCKE, M. *Adv. Morphogenesis*, **6**, (1967), 33–88 (cuticle structure and function)

NEVILLE, A. C. *Adv. Insect Physiol.*, **4**, (1967), 213–286 (nature and development of cuticular lamination)

RICHARDS, A. G. *The Integument of Arthropods*, Minneapolis, 1951

RUDALL, K. M. *Adv. Ins. Physiol.*, **1**, (1963), 257–313 (chitin-protein complexes in cuticle)

WIGGLESWORTH, V. B. *Ann. Rev. Entom.*, **2**, (1957), 37–54 (physiology of insect cuticle: review)

—— *The Principles of Insect Physiology*, 7th Edn., Chapman and Hall, London, 1972, 27–60 (insect integument: properties, formation and shedding of the cuticle)

2 Respiration

Terrestrial life is a perpetual conflict between the need for oxygen and the need for water; for those conditions which favour the entry of oxygen into the body favour also the escape of water. This conception, applied to the insects, illuminates the whole subject of their respiration. The majority of insects, as we have seen, have become encased in a waterproof covering; a covering which is relatively impermeable to oxygen also. They breathe, as was discovered by Malpighi (1669), through tracheal tubes, which open along the sides of the body through a series of spiracles, and convey air directly to the tissues (Fig. 7, A). The tracheal trunks anastomose one with another, and then branch and rebranch, becoming finer and finer, until their ultimate ramifications are either invaginated into the individual fibres of the muscles or the cells of other tissues, or form a network around them which is more or less rich according to their oxygen requirements.

The Tracheae and Tracheoles

Tracheal structure. Morphologically, the tracheae are invaginations of the ectoderm. Consequently, they consist of a cellular matrix and a lining cuticle. During its formation the epicuticle of the tracheae is thrown into regular spiral folds and the cells then secrete cuticular substance into these folds to form the spiral filaments or taenidia. This spiral formation enables the tracheal tubes to be bent and pressed upon without collapsing.

15

Chemistry. The chemistry of the tracheal cuticle resembles that of the general surface of the insect. The innermost layer is composed of the nonchitinous 'cuticulin', perhaps associated with wax; in the larger trunks a certain amount of chitin is present outside this; but in the small branches chitin is absent. The membrane lining the fine terminations is composed of polymerized but untanned lipoprotein.

Tracheoles. When the tracheal branches have become reduced to about 2 μm in diameter they break up abruptly to give rise to much finer tubes 0·6–0·8 μm in thickness. These are termed tracheoles. As viewed with the light microscope they appear to have smooth walls; but examination with the electron microscope has revealed that their lining membrane is thrown into spiral or annular folds just like the tracheae. The tracheoles run a course of some 200–400 μm gradually tapering to end in a blind rounded extremity at about 0·2–0·5 μm diameter. Each tracheole is really a single, greatly elongated cell with its nucleus lying about one-third of the way along it. Sometimes the tracheoles may be branched. It is possible that they may occasionally anastomose, but this is still uncertain. Often they arise from the side of quite large tracheae.

When the insect moults, new and larger tracheae are laid down by the tracheal epidermis, and the cuticular lining of all the tracheae is shed and drawn out through the spiracles. But the linings of the tracheoles are not shed: the old tracheoles are secured to the new tracheae by rings of a cementing substance resembling cuticulin.

Movements of water in tracheoles. The permeability of the tracheal system shows a gradation from the spiracle to the tracheole ending. To some extent it is permeable to gases throughout, though naturally it is much more so where the walls become so delicate as they are in the tracheoles. Like the epidermal cuticle, it is probably impervious to water, except in the tracheoles, which are freely permeable. Indeed, the terminal part of the tracheoles in many insects, both terrestrial and aquatic, normally contains a variable amount of fluid. There seems to be an equilibrium between the capillarity of the tracheoles (which are readily wetted by water) on the one hand and the 'swelling pressure' or 'colloid osmotic pressure' of the cytoplasm outside the tracheoles on the other. If the tracheoles are exposed artificially to slightly hypertonic fluids (1·0 per cent. sodium chloride), this will increase the 'swelling pressure' of the cells, fluid is sucked out of the tracheole endings and the column of air extends farther along the tracheoles towards the tissues. These same changes can be brought about by the increase in osmotic pressure which accompanies muscular contraction, especially if the insect is meanwhile exposed to air with a reduced content of oxygen (Fig. 5). This has the effect of improving the supply of oxygen locally to those tissues whose need is greatest. But in some insects the tracheoles seem always to contain air right to their extremities; perhaps in these cases the wetting properties of the tracheole membrane are different.

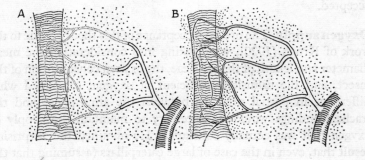

Figure 5. Tracheoles running to a muscle fibre: semischematic
A, muscle at rest; terminal parts of tracheoles (shown dotted) contain fluid; B, muscle fatigued; air extends far into tracheoles. (Modified after Wigglesworth)

These movements of fluid in the endings of the tracheoles are purely passive responses to simple physical forces around them. But the tracheoles are living cells and are responsible for maintaining the normal permeability of their lining membrane: as soon as the cells die fluid rapidly seeps into their endings. And when the insect first hatches from the egg, or at the time of moulting, the tracheole cells, and indeed the cells bounding the entire tracheal system, actively absorb the fluid contents.

The Diffusion Theory of Insect Respiration

The next problem in the physiology of the tracheal system is how oxygen is conveyed to the tissues. Although the tracheae are permeable to oxygen throughout their length, the abundant supply of tracheal capillaries in and around the most active tissues leaves no doubt that most of the oxygen enters the organs through these endings; and the problem resolves itself into how oxygen is conveyed from the spiracles to the tracheoles. It was suggested long ago by Treviranus (1816), and later by Thomas Graham (to whom we owe the laws of diffusion of gases), that it must be conveyed by diffusion. But to the superficial observer of these fine tracheal tubes, this idea seemed so inherently improbable that various other hypotheses were put forward, and it is only within comparatively recent years that the diffusion theory of insect respiration has been generally accepted.

Oxygen supply. The present acceptance of this theory is due to the work of Krogh (1920), who, taking into consideration the mean diameter and length of the tracheae, the oxygen consumption of the insect, and the diffusion coefficient of oxygen, calculated what difference in partial pressure between the atmosphere and the tracheal endings would be necessary to maintain the supply of oxygen that was actually consumed. He obtained the surprising result that, even in the case of large caterpillars (assuming that the spiracles remain permanently open), the partial pressure of oxygen at the tracheal endings need not be more than 2 or 3 per cent. below that in the atmosphere; showing that diffusion alone was adequate to supply the needs of the insect. The same is true of the tracheoles: even in the active flight muscles of insects, gaseous diffusion will account for all the oxygen which the cells require. But when the muscle fibres exceed about 20 μm in diameter the tracheoles indent the surface membrane and become 'internal'. It is calculated that in the locust *Schistocerca* the rate of diffusion of oxygen into the flight muscles via the tracheoles is about 100,000 times faster than it would be if all diffusion were in the liquid phase.

Carbon dioxide discharge. The same argument applies to the elimination of carbon dioxide. The amount of carbon dioxide produced by the insect is usually rather less than the oxygen taken in; their rates of diffusion, being proportional to the square roots of their densities ($\sqrt{22}:\sqrt{16}$), are not enormously different. Therefore, since diffusion alone will account for the supply of oxygen, it will account equally for the elimination of carbon dioxide. But carbon dioxide diffuses through animal tissues, and therefore presumably through the insect cuticle, more rapidly than oxygen. Consequently, since carbon dioxide liberated in the tissues must diffuse equally in all directions, the amount of this gas which will escape directly through the cuticle of the body wall (especially when this is thin) and through the walls of the larger tracheal trunks, must be far greater than the amount of oxygen that will enter by these routes. It is probably safe to say that in most terrestrial insects almost all the oxygen is taken in by the tracheae; in the case of *Carausius*, it has been shown that about 25 per cent. of the carbon dioxide escapes through the skin, and in larvae of *Dytiscus* and *Eristalis* somewhat less.

Function of the Spiracles

This simple conception of tracheal respiration is complicated by the need for retaining water; for most of the evaporation from an insect takes place through the spiracles. Burmeister (1832) discovered that each spiracle is provided with an occlusive mechanism, the main function of which is probably to protect the insect from loss of water: the adult flea, which has well-developed sphincters in the spiracles (Fig. 6, A), is far more resistant to desiccation than the larva which has none. The bug *Rhodnius* will survive for some weeks in a desiccator over sulphuric acid; but if it is caused to keep its spiracles open by the addition of 5 per cent. of carbon dioxide to the air it is completely dried up and dead within forty-eight hours (Fig. 6, B).

Spiracular movements. Normally the spiracles are kept closed, being opened only just often enough to keep the insect supplied with oxygen. In the flea, for example, while the insect is at rest all the respiration is effected through the first and eighth spiracles of the abdomen. These show a regular rhythm of opening and closing every five or ten seconds. If the insect struggles they remain open a little longer. If the rate of metabolism is increased by raising the temperature they open more frequently and other spiracles are brought into operation. And at the height of digestion and egg production in the female all the spiracles may be kept open for quite long periods. This necessarily results in an increased rate of water loss. We can see now how necessary it is for the tracheae to anastomose: while the spiracles are closed any organ or tissue which is active can draw oxygen from any part of the system; and the oxygen throughout the system can be renewed by the opening of a single pair of spiracles.

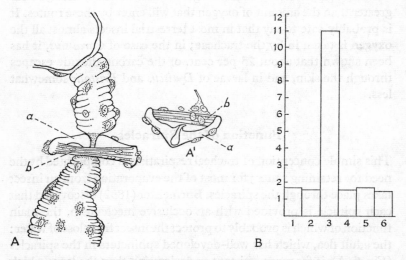

Figure 6. A, closing mechanism below an abdominal spiracle in the flea. A[1] the same seen in cross-section (after Wigglesworth). *a*, chitinous rod with a deep cleft enclosing the trachea; *b*, compressor muscle. B, daily loss of water in an adult *Rhodnius* starved in dry air. During the third day the spiracles were kept open by exposure to 5 per cent CO_2 (after Wigglesworth and Gillett). Ordinate: loss of weight in mg. per day. Abscissa: days

Control of spiracles. This function of the spiracles has been termed the 'diffusion control' of insect respiration. It is regulated by respiratory centres in the ganglia of the ventral nerve cord and in the brain. These centres are stimulated by carbon dioxide: exposure to 2 per cent. carbon dioxide causes the spiracles to remain permanently open. They are also stimulated by the acid metabolites which result from oxygen lack: traces of lactic acid injected into the flea will cause the spiracles to remain widely open even in the absence of carbon dioxide. In normal respiration oxygen want and carbon dioxide accumulation co-operate in the control. There appears to be some degree of control within the muscles of the spiracles themselves, for even after complete removal of the central nervous system weak responses remain.

Ventilation of the Tracheal System

This type of respiratory system, dependent solely upon gaseous diffusion, is adequate for most small insects, and even large insects if they are relatively sluggish; but it is insufficient for actively running or flying species with a high rate of metabolism and a massive demand for energy. In these, in the larger Orthoptera and beetles, in bees, wasps and flies, a mechanical ventilation of the tracheal system is superadded. We have seen that the spiral folding of the typical tracheae renders them resistant to collapse under pressure; but, in some cases at least, they can be expanded and contracted in their long axis like an accordion, and their capacity reduced by as much as 20 to 30 per cent. In other cases they are not round in cross-section but elliptical (*Dytiscus* larva) or even ribbon-like (in the thorax of Muscid flies) and will then readily collapse when the pressure around them is increased. Or they may have thin-walled cavities at intervals along their course (as in *Melolontha*), or large dilatations ('air sacs') which occupy a great part of the body cavity (as in many Orthoptera, Hymenoptera, &c.) (Fig. 7, B).

Air sacs. There is little doubt that, as Treviranus suggested, the main function of these collapsible tracheae or air sacs is the ventilation of the respiratory system. Like the ventilation of the lungs of vertebrates, the filling and emptying of these air sacs is secondary to the respiratory movements of the rigid body wall which encloses them. When the body wall is in the inspiratory position they are widely open; in the expiratory position many of them are collapsed and empty. In addition they are doubtless ventilated to some extent by general movements of the body – for example, during flight in the locust *Schistocerca*, automatic ventilation of the intermuscular air sacs is an essential element in the supply of oxygen; and where they lie in rigid portions of the body, such as the head, they are ventilated, as Graber (1877) pointed out, by the transmitted pressure of the blood.

Efficiency of ventilation. The result of this process is that the greater part of the tracheal system is kept filled with a gas which approximates in composition to the outside air; but the actual supply of oxygen to the tissues again takes place by diffusion along the tracheal branches given off by the air sacs. This, of course, is analogous to what happens in vertebrates, where only the upper parts of the respiratory tract are ventilated mechanically, exchanges in the alveoli of the lung being dependent on diffusion. The extent to which the tracheal system is ventilated mechanically varies enormously in different insects, and in the same insect according to its physiological state; but in many cases the efficiency of the process compares very favourably with the ventilation of the human lungs. For instance, in *Melolontha* the respiratory system may be emptied during expiration of about one-third of its total capacity (the total capacity being 39 per cent. of the body volume!), and in the larva of *Dytiscus* of about two thirds. (In quiet respiration in man, the lungs are emptied of about one-seventh of their total capacity; the most extreme degree of ventilation empties about two-thirds.)

Other functions of air sacs. In addition to their ventilating function, the tracheal air sacs of insects sometimes serve a purpose like that of the air-containing sinuses in the facial skeleton of mammals, or the pulmonary air sacs in the bones of birds; thus, by extending into the massive head and mandibles of certain beetles, they permit an increase in the skeletal bulk without adding to the insect's weight. They also allow for changes in size of the internal organs, such as the ovaries or the gut, without affecting the external form of the body. That is well seen in Muscid flies. And they serve to reduce the volume of the circulating blood and thus aid the circulation of sugar and other fuels to the active muscles.

Mechanism of ventilation. The respiratory movements of insects are brought about by a special musculature which varies greatly from one group to another. As a general rule the movements are confined to the abdomen, but in *Dytiscus* and *Hydrophilus* ventilation is maintained by aspirating movements of the metathorax, the abdominal movement apparently being passive. In most cases expiration is active and inspiration passive, though in the larva of *Aeschna* and in the grasshopper inspiratory muscles also are present. The movements may consist of dorso-ventral flattening movements (Orthoptera, Coleoptera) or longitudinal telescoping movements (Diptera, Hymenoptera).

The simplest interpretation of these movements is that they are ventilating all parts of the tracheal system alike. But if, during the respiratory contraction, the spiracular valves in different parts of the body open and close independently of one another, it is possible that a directed stream of air will be driven through the tracheal system – certain spiracles serving always for inspiration, others for expiration – an arrangement which would serve to enrich the air in the inspiratory section of the system. The idea that such a circulation of air does occur has often been put forward, but only in recent years has it been convincingly demonstrated. Thus, it has been shown that if the four anterior pairs of spiracles in the locust (*Schistocerca*) are enclosed in one gas chamber, and the six posterior pairs in another, air is actively transported during respiration from the anterior to the posterior chamber. The underlying principle of this mechanism is that the rhythmical nervous impulses which control the ventilating movements are co-ordinated with impulses which bring about the synchronous movement of the spiracles – some of these opening during inspiration, others during expiration. But before looking further into this point, we must consider the control of the ventilating movements themselves.

In some flying insects the pressure of the contracting muscles may serve as an alternating pump to drive air in and out of the tracheal system of the thorax. In *Schistocerca* the amount of air pumped through the thorax by the abdominal ventilation remains fairly constant at about 30 litres air/kg/h. But during level flight the thoracic pump provides an additional 250 litres and if need be its capacity can be increased up to at least 950 litres/kg/h.

The Regulation of Respiration

The respiratory movements are effected by impulses from nerve centres. These centres are of two kinds: so-called primary respiratory centres which lie in the segmental ganglia and control the movements of their own segments (for the isolated segments of the abdomen may perform respiratory movements), and secondary centres which have an overriding action and control the movements in the whole insect. One might expect the secondary centre to lie in the head, but this seems never to be the case: decapitation produces only a temporary and uncertain effect on respiration. In the few insects that have been studied, it is situated in one of the thoracic segments.

Respiratory centres. The isolated ganglia of insects which possess a mechanical ventilation of the tracheal system (*Dytiscus*, *Aeschna*) show rhythmical changes in electrical potential of the same frequency as the respiratory movements of the intact insect. There must, therefore, be a spontaneous respiratory rhythm inherent in the nerve centres. But the activity of the centres, both in rate of rhythm and depth of movement evoked, may be influenced by external stimuli, either nervous or chemical; and during times of rest the ventilating respiratory movements may cease altogether. There is evidence that during the onset of activity – flight, for example – the respiratory movements may sometimes be initiated by nervous stimuli alone (for any stimulus will cause an increase in respiration in insects); but they can also be initiated by chemical stimuli alone; and although there must be endless variation in different species, perhaps these play the greater part in the normal life of the insect.

Control of ventilation. Either a lack of oxygen or an excess of carbon dioxide may stimulate the respiratory centres and cause hyperpnoea; in some insects an excess of oxygen may cause a prolonged apnoea. Now excess of oxygen will not diminish the production of carbon dioxide; nor will a lack of oxygen increase it. If, therefore, carbon dioxide were the *sole* respiratory stimulant, respiration should not be influenced by the tension of oxygen. But there is no doubt that carbon dioxide *can* control respiration: in *Carausius*, the activity of the secondary or prothoracic centre is stimulated by carbon dioxide tensions of 0·2 to 3 per cent.; the primary centres respond to 12–15 per cent.; and in the *Periplaneta* and other insects, ventilation begins, even in the insect at rest, at a carbon dioxide tension of 10 per cent. In *Schistocerca* carbon dioxide is normally more important in control than oxygen lack. Since it is reasonable to assume that the chemical factor at work is the same in all these cases, it seems probable that this factor is the acidity in the nerve centres – due either to an excess of carbon dioxide, or to acid metabolites accumulating in the absence of oxygen.

Cyclic discharge of carbon dioxide. During the winter months the pupae of many insects go into a state of dormancy (p. 110): the rate of metabolism falls, and the respiratory centres become rather insensitive to carbon dioxide. In this state of dormancy large insect pupae show a curious cyclic discharge of carbon dioxide: oxygen uptake through the spiracles is more or less continuous, for oxygen want causes the spiracles to open at frequent intervals for very brief periods; but carbon dioxide output takes place in 'bursts' which may happen only once in twenty-four hours. At these times the large accumulation of carbon dioxide causes a prolonged opening of the spiracles. This phenomenon is, indeed, the result of the control of the opening and closure of the spiracles by the combined action of oxygen want and carbon dioxide accumulation in the blood – acting in large insects with a very low rate of metabolism.

We have seen (p. 19) that in thin-skinned insects much of the carbon dioxide diffuses through the general surface of the body, and therefore cannot serve as a respiratory stimulus at all. Thus we find that mosquito larvae, and the larvae of *Aeschna* and *Dytiscus*, are driven to seek air at the water surface by oxygen want and not by carbon dioxide excess. On the other hand, the leg movements in the

aquatic bug *Corixa*, which direct a stream of water over its air store
(p. 28), are called forth by the accumulation of carbon dioxide;
while the same insect is caused to rise to the surface when the air
store is reduced to a given size.

Directed streams of air. We are now in a position to revert to the
question of the stream of air directed through the tracheal system by
the co-operation of ventilatory and spiracular movements. We have
seen that in the insect at rest the supply of oxygen is effected solely
by diffusion through one or two pairs of spiracles. As the demand
for oxygen increases, more spiracles come into operation and they
remain open for longer periods. When the oxygen requirements are
still greater ventilatory pumping movements begin and then the
rhythm of opening and closure of the spiracles is changed in such a
way as to cause a directed stream of air through the system. All this
implies a most intricate nervous co-ordination between the control
of the pumping movements and the regulation of the spiracles; and
the rhythm may change completely as the intensity of stimulus
increases.

Regulation of Tracheal Supply

The general arrangement of tracheae is part of the inborn pattern
of growth in the insect, but this pattern is subject to modification in
detail in accordance with local demands for oxygen. (i) During
growth and moulting new tracheae and tracheoles grow out from the
existing system, and the new outgrowths are far more abundant in
regions with a deficient oxygen supply. (ii) In the intervals between
moults, of course, no new tracheae or tracheoles can be formed. But
if the epidermal cells have an inadequate supply of tracheoles they
give out elongate processes, sometimes as much as 100 μm long,
which become attached to remote air-filled tracheoles and then, by
contracting, draw these tracheoles towards them.

Respiration of Aquatic Insects

Breathing of atmospheric air. The majority of insects that are
aquatic in the adult stage breathe gaseous air like the terrestrial

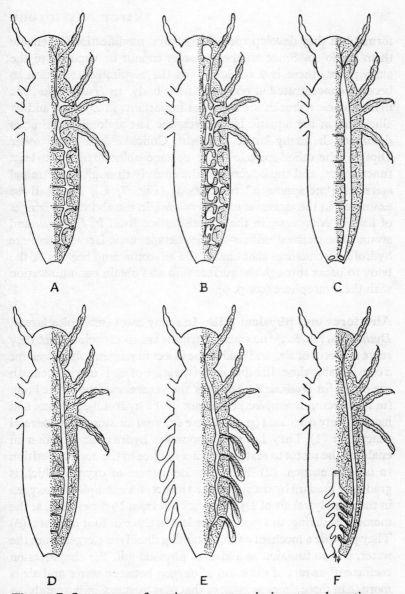

Figure 7. Some types of respiratory system in insects: schematic
A, simple anastomosing tracheae, with sphincters in the spiracles (p. 15); B, mechanically ventilated air sacs developed (p. 21); C. metapneustic respiration; only terminal spiracles functional (p. 28); D. tracheal system entirely closed: cutaneous respiration (p. 30); E, the same with abdominal tracheal gills (p. 29); F, the same with rectal tracheal gills (p. 30)

forms; but they develop special structural modifications to enable them to do so. Since all the spiracles cannot be exposed to the atmosphere, there is a tendency for the respiratory function to become concentrated at one end of the body. In *Hydrophilus*, the mesothoracic spiracles are the most important; in *Dytiscus* and its allies and in the aquatic Hemiptera, the last abdominal spiracles dominate. In many larvae (*Dytiscus*, Culicidae and many other Diptera) the other spiracles have become obliterated, or at least functionless, and the insects breathe entirely through the terminal spiracles ('metapneustic' respiration) (Fig. 7, C), In all these examples, in the antennae of *Hydrophilus*, in the abdominal fringes of hair in *Notonecta*, in the elytral surface itself in *Dytiscus*, and around the tracheal orifices of the metapneustic larvae, there are hydrofuge structures that enable the air-containing regions of the body to break through the surface film and obtain communication with the atmosphere (see p. 6).

Air stores and physical gills. In many cases (mosquito larvae, *Dytiscus* larvae, &c.) the tracheal trunks are so capacious that they serve as stores of air, and enable the insect to remain submerged for a considerable time. In other insects, a store of air is carried beneath the elytra (in *Dytiscus* &c.) or over the ventral surface of the body (in *Notonecta*, *Corixa*, &c.) by means of the hydrofuge surfaces and hairs already discussed (p. 6). These external air stores have several functions: (1) They have an important hydrostatic function in enabling the insect to reach the water surface in the correct position to take in oxygen. (2) They provide a store of oxygen which is gradually used up by the submerged insect; for example, the oxygen in the sub-elytral air of *Dytiscus* may fall from 19·5 per cent. at the moment of diving, to 1 per cent. or less in three or four minutes. (3) They provide a mechanism for obtaining dissolved oxygen from the water, and so function as a kind of physical gill. For the 'invasion coefficient' or rate of diffusion of oxygen between water and air is more than three times as great as that of nitrogen. Consequently, as the partial pressure of oxygen in the air store becomes reduced, equilibrium will be restored by the diffusion of oxygen inwards, rather than by the diffusion of nitrogen outwards. Of course, some nitrogen will diffuse out, but as long as any remains undissolved the process can go on, and the insect can extract dissolved oxygen from the water. This mechanism is of more or less value to all aquatic insects that carry air stores; and small forms like *Corixa* can obtain enough oxygen in this way even at summer temperature so long as they do not swim actively. The striking effect of the process may be shown by quoting a single experiment: *Notonecta* lived five minutes submerged in water saturated with nitrogen, thirty-five minutes in water saturated with oxygen, but six hours in water saturated with atmospheric air – the air store, in each instance, being first charged with the dissolved gas.

Plastron respiration. In a few insects, such as the beetles *Haemonia* and *Elmis* and the Naucorid bug *Aphelocheirus*, the hydrofuge hairs which carry the film of air are bent over at the tip so as to provide a hydrophile surface enclosing a layer of air which cannot be replaced by water (Fig. 8). This firmly held layer of gas is termed a 'plastron'; it enables such insects to become independent of the atmospheric air and to obtain all their oxygen from the water.

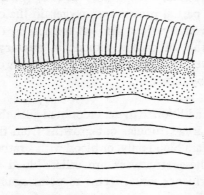

Figure 8. Section through the cuticle of an abdominal sternite of *Aphelocheirus* showing plastron hairs, each about 5 μm long, bent over at the tip (modified, after Thorpe and Crisp)

Tracheal gills. A much more common mechanism, by which the insect has become completely adapted for the respiration of dissolved oxygen, is the obliteration of all the spiracles and the development of tracheal gills. In many cases, the general surface of the larva, richly supplied with fine tracheae, provides the respiratory surface (in *Chironomus, Simulium, Corixa,* &c.) (Fig. 7, D). Sometimes there are specialized plate-like projections or filaments rich in tracheae (Ephemeroptera, Plecoptera (Fig. 7, E); or papillae from the anus, consisting of little more than massed tracheal branches covered by a thin cuticle (*Elmis,* Col.); or a network of tracheae inside the rectum (as in the larvae of Anisopterid dragon-flies) (Fig. 7, F); or rigid cuticular filaments containing air (cuticular gills). The air spaces in cuticular gills are in direct contact with the water through numerous perforations; they thus form an example of plastron respiration. They do not collapse on drying, and can therefore serve for respiration both in water and on land (as in the pupa of *Simulium,* and other insects from mountain torrents).

In some cases the oxygenation of these organs is ensured by a highly developed respiratory centre; the gill plates of Ephemeroptera are violently agitated in water poor in oxygen, and the rectum of *Aeschna* is ventilated by abdominal movements controlled by primary and secondary respiratory centres (p. 24). It has been shown that the partial pressure of oxygen in the closed system of these insects is always less than its tension in the surrounding medium, and that of carbon dioxide is greater; it is therefore generally believed that the exchange of gases through the gills is effected solely by diffusion.

Blood gills. In addition to these undoubted gills, there are, in many aquatic insect larvae, tubular out-growths from the body surface which may contain blood only, or blood with a rather sparse supply of tracheae. These structures are called blood gills; but what experimental evidence exists seems to point against their having any but a subsidiary function in respiration. In one case, that of the anal papillae of the mosquito larva, they are concerned in the active uptake of ions (sodium, potassium and chloride) from the surrounding water and thus play an important part in nutrition.

Hydrostatic organs. We have seen (p. 28) that the external air stores carried by aquatic insects help to control their equilibrium in the water. The air in the tracheal system, also, must often be of some importance for this purpose; and in the pelagic larva of *Corethra* there are developed four large air sacs which function solely as hydrostatic organs. This larva can adjust its specific gravity to that of the water it is in by varying the capacity of these air sacs – apparently by some active change which causes expansion or contraction of their cuticular walls.

Respiration of Parasites

The internal parasites of insects, which live a semi-aquatic life within the blood and tissues of their host, naturally show many of the respiratory adaptations possessed by aquatic larvae. In the young stages of many of them the tracheae do not contain air, and oxygen simply diffuses from the blood of the host into the blood of the parasite – as it does in the young aquatic larvae of *Chironomus* (Dipt.) and *Acentropus* (Lep.), &c. Later, when air has appeared in the tracheae, these commonly supply a rich network of branches to the skin – as they do in the older aquatic larvae of *Simulium*, *Chironomus*, &c. There are often outgrowths from the body surface, usually from the tail, which recall the gill-like organs of aquatic forms. Sometimes these structures are well supplied with tracheae (as in the larva of *Cryptochaetum* (Agromyzidae: Dipt.)) or with circulating blood (as in *Apanteles* (Braconidae: Hym.)) and are of proved importance in respiration; but in most cases their respiratory significance is very doubtful and their function is problematical. Finally, the metapneustic type of respiration common in aquatic insects is seen in most Dipterous parasites (Tachinidae, &c.), which pierce the body-wall or the large tracheal trunks and breathe the atmospheric air. The habit of attaching themselves to the tracheal tubes recalls the fact that certain aquatic larvae (of the beetle *Donacia* and the mosquito *Mansonia*) possess specially modified respiratory siphons which they insert into the air-containing tissues of aquatic plants, and are thus able to remain permanently beneath the water surface.

The role of haemoglobin in insect respiration is discussed on p. 42.

BIBLIOGRAPHY

HINTON, H. E. *Adv. Insect Physiol.*, **5**, (1968), 65–162 (spiracular gills: review)

MILLER, P. L. *Adv. Insect Physiol.*, **3**, (1966), 279–354 (regulation of insect respiration: review)

THORPE, W. H. *Biol. Rev.*, **25**, (1950), 344–390 (plastron respiration: review)

WEIS-FOGH, T. *J. Exp. Biol.*, **41**, (1964), 229–256; **47**, (1967) 561–587 (ventilation and diffusion in locust respiration)

WIGGLESWORTH, V. B. *Biol. Rev.*, **6**, (1931) 181–220 (insect respiration: historical review)

—— *The Principles of Insect Physiology*, 7th Edn. Chapman and Hall, London, 1972, 357–410 (insect respiration)

3 The Circulatory System and Associated Tissues

In the insect body there is only one tissue fluid, the blood or haemolymph, occupying a single cavity, the haemocoel, and separated from the tissue cells only by the delicate but continuous basement membranes. With few exceptions, there is only one blood-vessel, which runs along the midline of the back. The posterior segment of this pulsating vessel, the 'heart', is provided with a series of valved openings or ostia through which the blood can enter; the anterior segment, the 'aorta', is a uniform contractile tube. After passing through the brain above the oesophagus, the aorta ends more or less abruptly; and from this point the blood simply percolates slowly backwards through the tissues. But in many insects it is still subject to some direction: the aorta may discharge into definite vessels carrying the blood in different directions; the antennae and limbs are often divided by longitudinal membranes, the blood entering the limb on one side and leaving on the other; and the abdominal cavity in some insects is divided by two horizontal membranes, the dorsal and ventral diaphragms, into three sinuses: the pericardial sinus containing the heart, the perineural sinus containing the nerve cord, and the visceral sinus between. The perineural sinus tends to conduct the blood to the posterior end of the abdomen before allowing it to return to the heart; but it is perforated and incomplete laterally, so that some of the blood circulates transversely across the abdomen below it. The dorsal diaphragm, when present, is also fenestrated, and thus allows the blood to enter the pericardial sinus throughout its length (Fig. 9).

Mechanism of the Circulation

Blood pressure. In insects with a rigid cuticle, the blood in the body cavity, when the insect is at rest, is at atmospheric pressure (Odonata) or even less (*Dytiscus, Apis*). In the more soft-bodied insects, the general pressure may be raised (to 40 or 50 mm of water in the larva of *Aeschna,* or even more during exertion) by the muscular tension of the body-wall. The pressure may be increased, also, during the ventilation of the tracheal system; and during those special acts which are brought about by the displacement of the blood, such as the expansion of the wings in emerging adults (p. 10), the general pressure may be maintained for a prolonged period at a high level (75 mm of water in the dragon-fly).

Propulsion of the blood. But, like the blood in the vena cava of mammals, the blood of insects is always aspirated into the heart, during diastole, under a negative pressure: a pressure less than that existing in the general body cavity. This force is due partly to the elastic muscular walls of the heart itself, partly to the elastic traction of the dorsal diaphragm which is tied to the lower wall of the heart, and partly to the contraction of the muscle fibres (the aliform muscles) which commonly occur in the substance of this diaphragm. During systole, a weak positive pressure develops and the blood is driven forwards.

In some insects, such as the larvae of Nematocera and the Ephemeroptera, the valves of the ostia have become so modified as to form interventricular valves; these divide the heart into a succession of chambers through which the blood can flow only forwards. But this arrangement is the exception; in most insects the heart is patent throughout its length and acts as a whole, the blood being carried along by peristaltic waves. As a rule, such waves start at the hind end of the heart and pass forwards; but in many insects the direction of beat may be periodically reversed, and the blood then escapes into the abdominal cavity through the relaxed ostia. Sometimes beats may arise at a number of points simultaneously and extinguish one another when they meet.

Accessory pumps. Thus the circulation of the blood is secured primarily by the work of the heart, which aspirates it from the abdominal cavity and pumps it forwards to the head. In the thorax, accessory pumps are often present. These aspirate blood from the thoracic cavity, through certain of the wing veins, and return it through connecting vessels either to the aorta itself (*Dytiscus*, *Aeschna* larva, *Sphinx*) (Fig. 9, A) or to the body cavity again (*Tabanus*, Dipt.; *Vespa*, *Apis*, Hym.). Accessory pumping organs may irrigate the antennae (*Periplaneta*, Hymenoptera, &c.), the legs (Hemiptera), and the wings (Diptera). But in many insects these structures are wanting, and the circulation through the appendages

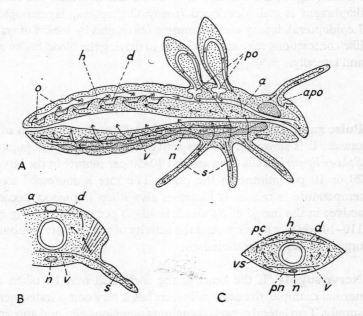

Figure 9. A, insect with fully developed circulatory system: schematic; B, transverse section of thorax of the same; C, transverse section of abdomen. Arrows indicate course of circulation. (Based largely on Brocher)

a, aorta; *apo*, accessory pulsatile organ of antenna; *d*, dorsal diaphragm with aliform muscles; *h*, heart; *n*, nerve cord; *o*, ostia; *pc*, pericardial sinus; *pn*, perineural sinus; *po*, meso- and metathoracic pulsatile organs; *s*, septa dividing appendages; *v*, ventral diaphragm; *vs*, visceral sinus is then effected (i) partly by their own muscular movements, (ii) partly by the pressure changes in the abdomen brought about by the respiratory movements (*Aeschna* larva, Ephemeroptera larvae) – in which case the pulse-beats in the leg synchronize not with the heart-beat but with the respiratory contractions, and (iii) partly owing to the fact that the efferent stream (in the posterior chamber of the limb) communicates with the perineural sinus, where the pressure is higher, while the afferent stream (in the anterior chamber) communicates with the perivisceral sinus, where the pressure is lower: the blood in the limb is thus subject to the aspirating action of both the heart and of the pulsatile organs of the thorax (Fig. 9, B). Finally, in those insects in which the ventral diaphragm is well developed (many Orthoptera, Hymenoptera, Lepidoptera), it may contain muscle fibres, and by means of wave-like contractions may actively assist in driving the blood backwards and laterally.

Control of the Heart-beat

Pulse rate. The pulse-rate of the dorsal vessel varies from many causes. It is slower in the less-active stages, such as the pupa: in *Sphinx ligustri* it falls from around 40–50 per minute in the larva to 20 or 10 per minute in the pupa. The rate is increased as the temperature is raised. It increases also when the insect becomes active: in the imago of *Sphinx* it is 40–50 per minute during rest, 110–140 during activity. And the activity of the accessory pulsating organs shows similar variations.

Nerve supply. If the heart of the cockroach may be taken as a normal example, the heart of insects has a very complicated nerve-supply. Two lateral nerves, containing ganglion cells, and apparently constituting an intrinsic cardiac ganglion, run along its entire length and are connected in front with the ganglia of the visceral nervous system. In addition, the heart receives sensory and motor nerve-fibres from the segmental ganglia of the ventral chain. There is no doubt, therefore, that the heart enjoys a large degree of nervous control; experimentally, faradic stimulation in the region of the neck gives rise to accelerating impulses which reach the heart both by the lateral nerves and by the segmental branches from the ventral cord. The heart is also supplied with neurosecretory axons (p. 155); some of these have cell bodies in the ventral ganglia, others in the lateral nerve cords of the heart. The function of their secretion is not known.

Myogenic rhythm and hormonal control. But though influenced by the nerve-supply in this way the property of rhythmical contraction resides in the heart-muscle; the heart-beat is 'myogenic'. For in many insects, such as *Aeschna*, ganglion cells are absent, and yet the isolated heart or even fragments of the heart continue to beat rhythmically. This rhythmical activity is greatly influenced by the normal tension exerted upon the heart by the alary muscles. But it is probably true to say that the pace-maker of the heart is neurogenic; and this pace-maker is affected by acetylcholine, which causes acceleration. Adrenaline, which appears to act upon the post-ganglionic fibres, also causes acceleration; and under the stimulus of feeding the corpus cardiacum of *Periplaneta* liberates a peptide hormone which causes the pericardial cells to secrete a cardiac stimulator, perhaps an indolalkylamine related to serotonin, which increases the rate and amplitude of the heart beat.

Composition of the Blood

Physical properties. The blood or haemolymph of insects may be a clear, colourless fluid; but more often it is tinged with green or yellow (p. 91). The specific gravity varies from $1 \cdot 03$ in the larva of *Celerio* (Lep.) to $1 \cdot 045$ in the larva of *Apis*. The reaction is usually very slightly acid: $pH = 6 \cdot 4$ in *Celerio* larva, $6 \cdot 6$–$6 \cdot 8$ in *Bombyx mori*, $6 \cdot 83$ in *Apis* larva; but it may differ slightly in the two sexes and may vary with the age of the insect. Thus, in *Bombyx* it becomes more alkaline at each moult; and in *Apis*, *Celerio* and other insects it is more acid during the pupal stage. The haemolymph is moderately well buffered; bicarbonates, phosphates, amino-acids and, chiefly, proteins are responsible. The total molecular concentration is rather high. Expressed in terms of sodium chloride concentrations of the same osmotic pressure, we get instead of $0 \cdot 9$ per cent., the usual value in mammals, such values as $1 \cdot 5$ per cent. in the bee larva, $1 \cdot 38$ in *Pieris* larva, $2 \cdot 12$ in the larva of *Tenebrio*; but lower values occur in aquatic insects, such as $0 \cdot 75 – 0 \cdot 85$ in the larvae of mosquitos, $0 \cdot 69 – 1 \cdot 0$ in Ephemeroptera.

Ionic composition. The ionic composition is exceedingly diverse in different groups of insects. Among exopterygotes the concentration of chlorides may approximate to that found in mammals (65–70 per cent. of the total molecular concentration) but in many insects chlorides contribute no more than 15 per cent. On the other hand, the concentrations of phosphate, calcium and magnesium are relatively high as compared with vertebrate blood. A very curious feature of insect blood is the extreme variability of the ratio of sodium to potassium in different species. In many insects the potassium concentration is so high that nerve-fibres will not conduct impulses when exposed to it. In such insects the nerves continue to function only because the perineurium cells below the fibrous sheath that covers the nervous system actively maintain around the axons a tissue fluid with a very low potassium content.

Organic chemistry. Another striking feature is the high content of amino acids, which may be twenty or thirty times as great as in human blood and form nearly 15 per cent. of the total nitrogen. In addition, there is a considerable amount of residual nitrogen in peptide form. Proteins may be present in much the same amounts as in mammalian plasma (around 6 per cent.) but usually there is much less than this. They can be grouped as 'albumens' and 'globulins', but by the use of electrophoresis and antigen – antibody precipitation as many as twenty proteins can often be distinguished. Many of these are conjugated with lipids, sterols, or carbohydrates and most of them are enzymes of one kind or another. Lipids are usually transported by the blood in the form of lipo-proteins, commonly carrying diglycerides, sometimes fatty acids.

In the honey-bee and some other insects the chief sugar of insect blood is glucose, but in most insects there is almost no fermentable sugar in the haemolymph; the chief carbohydrate is trehalose, a nonreducing glucose-glucose disaccharide which may sometimes be present at a concentration exceeding 5 per cent. A great variety of organic metabolites (such as succinate, pyruvate, citrate, &c.) and notably organic phosphates (such as α-glycerophosphate, glucose-6-phosphate, &c.) are present in insect haemolymph at levels that occur in vertebrates only within the tissue cells. Whereas fermentable sugars are often absent from the haemolymph, other reducing substances (perhaps phenols) are plentiful.

Cytology of the Blood

Plasmatocytes. The blood-cells or haemocytes of insects, as seen in preparations, present an extraordinary diversity of appearance; but this seems to be due to the protean forms which they can assume rather than to the abundance of distinct types. They multiply and grow in the body cavity throughout the life of the insect, appearing first as small darkly staining forms termed 'proleucocytes', which are often seen dividing and are not yet capable of phagocytosis. As they grow they become pyriform or spindle-shaped and will ingest dead bacteria, Indian ink, tissue debris, &c.; they are than called 'plasmatocytes'. Their appearance naturally varies very much with their content, and sometimes separate names have been given to these phases; moreover, they have a habit of spreading themselves out in stellate form upon flat surfaces, such as the basement membranes, their aspect becoming totally changed.

These haemocytes are by no means all circulating; indeed, the majority adhere to the surface of the tissues; they collect particularly along the sides of the dorsal vessel, often in definite clumps termed 'phagocytic organs'. The larvae of Chironomidae present a complete series in this respect: in certain genera only free haemocytes are present; in others, both haemocytes and fixed phagocytic cells occur; and in yet others there are no circulating cells at all but only phagocytic tissue. Many so-called phagocytic organs are in fact 'haemopoietic organs' concerned in producing and liberating new haemocytes within the circulating blood. Such organs are conspicuous along the dorsal vessel in Orthoptera; the 'lymph glands' of Drosophilia are another example.

Function of plasmatocytes. The haemocytes become far more numerous during moulting and metamorphosis, when they may play a part in removing the dead cells and tissues: during the metamorphosis of Muscids, these phagocytic cells (the 'Körnchen-kugeln' of Weismann), stuffed with the products of histolysis, are exceedingly conspicuous. The phagocytic haemocytes will ingest, and in some cases destroy, living bacteria introduced into the blood. They collect at the site of wounds and form a closing plug which forms the basis for the subsequent healing process. They will congregate in great numbers around foreign bodies and certain parasites, walling these off in a discrete capsule; sometimes the cellular nature of this capsule persists, but more often the cell bodies become converted into homogeneous membranes from which the nuclei disappear. The plasmatocytes often contain abundant inclusions of neutral mucopolysaccharide which are discharged when the cells apply themselves to the surface of the tissues during moulting. They probably contribute to the connective tissue membranes. In some insects, haemocytes laden with conspicuous granules (granular leucocytes) are present in the blood; although these are no longer phagocytic they seem to represent a stage in the life history of these same cells.

Oenocytoids. Besides this type of cell with all its varied powers, there is, in most insects, another type which is equally distinct. It has a rounded or oval form, the cytoplasm is generally oesinophil, and it takes no part in phagocytosis nor in the formation of encapsulating membranes. In appearance these cells are like diminutive oenocytes; and on this account they are often called 'oenocytoids'; but they do not seem to bear any relation to the true oenocytes. Under phase contrast they often show conspicuous inclusions and are therefore sometimes called 'granular cells'. Their function is quite unknown: they increase in number during moulting, like the plasmatocytes; and they collect around foreign bodies in a zone peripheral to the plasmatocytes themselves.

Finally, in certain Hemiptera, there are blood-cells laden with fat globules, 'adipocytes', and others charged with wax; but the relation of these to the types described above is uncertain.

Functions of the Blood

Mechanical. The blood plays an important part in transmitting pressure from one region of the body to another: the hatching of many insects from the egg (p. 11), the rupture of the old skin at moulting (p. 10), the expansion of the wings in the newly hatched adult insect (p. 11) and many other movements are brought about by the localized pressure of the blood.

In blood coagulation. In some insects, such as the larva of the honey-bee and in *Rhodnius* and many other Hemiptera, the blood does not clot; but in most insects it does, and thus plays an important part in the closure of wounds. In the haemolymph of most insects, besides the haemocytes mentioned above, there are certain 'hyaline haemocytes', indistinguishable in ordinary stained preparations but easily recognized with the phase contrast microscope. When the blood coagulates these cells extrude thread-like pseudopodial expansions or produce a fine precipitate in the plasma around them.

In connective tissue formation. The neutral mucopolysaccharide inclusions of the plasmatocytes are discharged during the course of moulting on to the basement membranes and connective tissues which separate the tissue cells from the haemolymph. What proportion of these membranes is contributed by the blood cells and what proportion by the tissue cells themselves is uncertain and surely varies from one tissue to another.

In nutrition. The blood conveys the nutrient materials and hormones to the tissues, and the waste products to the excretory organs. The precise part that it plays in these functions is not really understood, but it is probable that the composition of the blood may vary enormously with the state of nutrition and, notably, during moulting and metamorphosis when many of the larval tissues are breaking down and the cells of the fat body are yielding up their protein and fatty contents. The blood must also be regarded as an important reserve of food material: in the larva of *Deilephila*, during fasting, the protein in the blood is rapidly consumed; and during pupal life in the same insect more than half the total energy metabolism is effected at the expense of the blood.

In metabolism. The haemocytes may also play a part in intermediary metabolism; for example, in the conversion of tyrosine to the polyphenols needed for melanin and sclerotin formation, and perhaps in the formation of some of the blood proteins.

In immunity. The phagocytic blood-cells provide perhaps the chief mechanism for protecting the insect from bacterial invasion; but humoral immunity, both natural and acquired, does occur. The antibacterial factor in the blood and in the gut of insects appears to be a 'lysozyme' (p. 56). The full discussion of these matters would take us too far afield.

In respiration. We have already seen that oxygen is ordinarily conveyed directly to the tissues by the tracheal system. But many cells are separated from the nearest tracheal tubes by an appreciable space; sometimes, notably in the pupa, organs may be entirely devoid of tracheal supply; and in some aquatic insects the tracheal system is completely filled with fluid. Moreover, the tracheal system, throughout its length, is permeable to oxygen, which must therefore escape into the blood. Under all these circumstances, the blood acts as a carrier of oxygen. There are, also, here and there among insects, special arrangements of the tracheal system which seem to be designed to aerate the blood; a convoluted tract in the aorta of the honey-bee is richly supplied with tracheae; the posterior region of the heart in the metapneustic larvae of Nematocera is often surrounded by a basket-work of tracheoles; and in the thorax of *Nepa* is a peculiar organ so richly supplied with tracheae and bathed with blood as to suggest that it may be a kind of tracheal lung.

Haemoglobin. In view of these relations the question naturally arises whether the blood of insects contains chemical carriers of oxygen comparable with the haemoglobins and haemocyanins of other animals. With a single exception, this has not yet been proved to be the case. In those insects such as the larva of the honey-bee, in which the properties of the blood have been most carefully studied, it has not been found to take up more oxygen than can be accounted for by physical solution; but it must be admitted that such investigations have not yet been very extensive. The single exception is in the larvae of certain Chironomidae, which contain haemoglobin in the blood in free solution – that is, not in corpuscles. The affinity of this haemoglobin for oxygen is, however, very different from that of mammals. At all ordinary tensions of oxygen it remains fully saturated; and it only begins to liberate its oxygen when the pressure of this gas is reduced to 1 per cent. of an atmosphere or less; the precise level varying in different species. In other words, it is capable of acting as a carrier only under conditions of extreme paucity of oxygen and seems, in fact, to be a special adaptation to life in poorly oxygenated waters – the larvae that contain haemoglobin being much less susceptible to oxygen want.

We have seen, also, that the blood probably plays a considerable part in the carriage of carbon dioxide from the tissues (p. 18). But here again there seems to be no chemical provision for its transport; the bicarbonate and the carbon dioxide capacity of insect blood are both very low.

Organs and Tissues associated with the Blood

The blood itself is an important tissue; but it is not commonly recognized as such because the cells of which it is composed are free and unattached and, with the exception of the 'phagocytic organs' (p. 39), show no constant arrangement. But there are other organs and tissues which are likewise bathed by the circulatory fluid and which perform their functions solely through exchanges with it.

Pericardial cells are scattered along the heart, alary muscles and aorta. These have been considered in the past to have an excretory function (see 'nephrocytes' p. 63); but nowadays they are regarded as comparable with the reticulo-endothelial system of vertebrates (p. 64). We have already referred to them as a source of a cardiac accelerator (p. 37).

The fat body is often the most conspicuous object in the body-cavity of the insect. It consists of a loose meshwork of lobes, invested in delicate connective-tissue membranes, so as to expose the maximum of surface to the blood. The whole mass has a fairly regular anatomical arrangement constant in each species. The fat body is important in the storage of reserves of fat, glycogen and protein (p. 78) and sometimes of uric acid and lime (p. 65). But it has also important functions in intermediary metabolism. The cells are stuffed with mitochondria. They are almost as rich in diverse enzymes as the mammalian liver: esterase (lipase) to liberate fatty acids from the stores of fat; succinoxidase and all the other enzymes of the citric acid cycle which furnish hydrogen as fuel for the cytochrome system; transaminases which convert one amino acid to another; enzymes to convert glucose to trehalose; enzymes concerned in diamination, uric acid synthesis, purine oxidation and many more. They contain much ribonucleic acid, and synthesize protein for the circulating blood and for supply to the developing eggs. The conspicuous proteinaceous inclusions in the pupal fat body of many holometabolous insects seem largely to be proteins sequestered from the haemolymph. In some insects, notably in the larva of *Gasterophilus*, certain fat body cells (so called 'tracheal cells') become laden with haemoglobin; in the aquatic bugs *Buenoa* and *Anisops* the haemoglobin in similar cells serves as a store of oxygen during diving.

Oenocytes. Unlike the other tissues considered in this chapter, all of which are of mesodermal origin, the oenocytes are derived from the ectoderm. They often arise throughout the larval stages, for example in *Rhodnius*, from the ordinary epidermal cells. They may remain between the epidermis and basement membrane, or they may become distributed throughout the fat body. These cells also are probably important in intermediary metabolism. The only suggested function for them for which there is positive evidence is that they are ectodermal cells which have become specialized for the production of some particular constituent of the cuticle (and perhaps of the egg-shell) such as the lipo-protein of the cuticular layer or the cuticular wax (p. 5). There is some evidence that they are specifically concerned in the synthesis of hydrocarbons. Cytologically they are characterized by the presence of abundant smooth surfaced endoplasmic reticulum.

BIBLIOGRAPHY

ASHHURST, D. F. *Ann. Rev. Entom.*, **13**, (1968), 45–74 (connective tissue of insects: review)

FLORKIN, M. and JEUNIAUX, C. *Physiology of Insecta*, III (Morris Rockstein, Ed.), Academic Press, New York, 1964, 110–152 (composition of the haemolymph)

GREGOIRE, C. *Physiology of Insecta*, III (Morris Rockstein, Ed.), Academic Press, New York, 1964, 153–189 (coagulation of the haemolymph)

KILBY, B. A. *Adv. Ins. Physiol.*, **1**, (1963), 111–174 (biochemistry of insect fat body: review)

SALT, G. *The Cellular Defence Reactions of Insects*, Cambridge University Press, London, 1970, 118 pp

SHAW, J. and STOBBART, R. H. *Adv. Ins. Physiol.* **1**, (1963), 315–399 (osmotic and ionic regulation in insects: review)

WIGGLESWORTH, V. B. *Ann. Rev. Entom.*, **4**, (1959), 1–16 (insect blood cells: review)

—— *J. Reticuloendothelial Soc.*, **7**, (1970) 208–216 (pericardial cells: review)

—— *The Principles of Insect Physiology*, 7th Edn., Chapman and Hall, London, 1972, 411–475 (circulatory system, composition of blood, haemocytes, fat body, pericardial cells, oenocytes, etc.)

WYATT, G. R. *Ann. Rev. Entom.*, **6**, (1961), 75–102 (biochemistry of insect haemolymph: review)

4 Digestion

Insects feed upon almost every type of organic substance found in nature: some on plants, others on animals; some on the sap of plants or the tissue fluids of animals, others on foliage, or flesh, dry timber, or hair and feathers; some on fungi or the live and dead bacteria in the excrement of animals, others on the sterile juices of living forms. Their feeding mechanisms, and the structure and chemistry of their digestive system, present, therefore, the most extraordinary variety, and it becomes singularly difficult to sift out the general principles in the physiology of their nutrition.

The Alimentary Canal

Anatomy. With the exception of the earliest stages of some parasitic insects, which absorb nutriment through the general body surface, all insects take their food into an alimentary canal. This consists always of three parts: the fore-gut, mid-gut and hind-gut. The fore-gut and hind-gut are both lined with cuticle; in the mid-gut the cells are freely exposed; from which it follows that secretion of digestive juices can occur only in the mid-gut, and here, also, at least the greater part of absorption undoubtedly takes place. In many insects the secretion of the mid-gut is supplemented by that from the salivary glands, mixed with the food before it is swallowed.

Fore-gut and crop. Fig. 10 shows some of the commoner types of intestinal system, and the different uses to which their parts are put. In the most primitive insects, and many larval forms (Diptera-

Nematocera, Lepidoptera, Tenthredinidae, and many Coleoptera) (Fig. 10, A), the fore-gut or oesophagus has no other function than to conduct the food into the mid-gut; from which it is passed on, often more or less continuously, to the hind-gut. But in very many insects (Dermoptera, Orthoptera, Isoptera, Odonata, Hymenoptera, many Coleoptera) the hinder part of the fore-gut is dilated to form a capacious crop (Fig. 10, B). Here the food is stored before being transmitted, in small quantities at a time, to the mid-gut. It is, however, not only stored but digested, being acted upon always by the salivary secretion, and, at least in the Orthoptera and many Coleoptera, by the digestive juices passed forward from the mid-gut. In the higher Diptera a crop for storing food is again present (Fig. 10, C), but it takes the form of a diverticulum connected to the fore-gut by a narrow duct. The food in this crop is mixed only with saliva, and it suffers very little digestion until it is transferred to the long coiled mid-gut. In the Lepidoptera, and in many Diptera (Culicidae, Tabanidae) (Fig. 10, D), the ingested food goes straight to the mid-gut and is there both stored and digested; the 'crop' has all but lost its function as a food reservoir, and is used chiefly (in many Lepidoptera, solely) to receive the air swallowed by the insect when it emerges from the pupa (p. 10). In the fleas (Siphonaptera) and sucking lice (Siphunculata) the crop has been lost entirely (Fig. 10, E), and the food is taken straight into a voluminous stomach, where it remains until digestion is complete.

In some insects the crop seems to have a protective function: fluids with a high osmotic pressure, which is detected by special sense organs, are retained in the crop and passed only very gradually to the mid-gut; in mosquitos, whereas blood is swallowed directly to the mid-gut, syrup is deflected temporarily to the crop and passed on to the mid-gut very slowly.

The mid-gut. In many Diptera the mid-gut consists of several segments characterized by differences in the epithelium; the more anterior being concerned probably only with absorption, the more posterior with digestion and absorption. The value of this arrange-ment when the insect takes in much fluid with its food is obvious; for much of the water and many of the assimilable constituents are

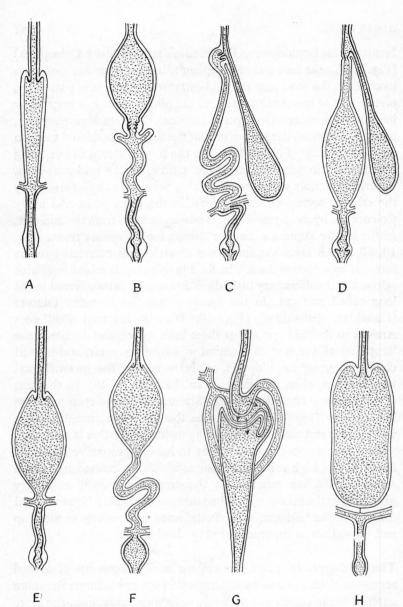

A   B   C   D

E   F   G   H

Figure 10. Types of alimentary system in insects: schematic
Fore-gut and hind-gut indicated by a heavy line internally. Explanation in text.
(G modified after Weber)

removed before the food reaches the digestive juices, and consequently these juices do not suffer unnecessary dilution. In many Heteroptera, this arrangement has gone a stage further; there is once again a capacious crop in which the meal is received, but it is a crop composed of mid-gut. In the blood-sucking forms (*Cimex* and *Rhodnius*) (Fig. 10, F), the food is not at all digested in this part of the mid-gut, but is merely concentrated by the removal of fluid, and the concentrated product passed on for digestion to the long narrow intestine.

The filter system. In these insects the fluid that is absorbed passes into the blood and is then excreted by the Malpighian tubes; but an advance on this mechanism has been developed by many Homoptera (Cicadoidea, Coccoidea, &c.). These feed on the juices of plants, and therefore receive a great excess both of water and of sugars. But instead of the superfluous fluid being taken into the blood and then eliminated by the Malpighian tubules in the manner just described, a dilated loop of the mid-gut at its commencement, with very delicate walls, is intimately associated with the posterior end of the mid-gut. In this way the fluid contents of the food can be absorbed at the beginning of the mid-gut and transferred directly to the rectum without the haemolymph or the general content of the mid-gut being unnecessarily diluted. There are many modifications of this so-called 'filter complex' and much remains to be learned about its physiology. The excess fluids are discharged to the exterior as honey-dew, &c. (p. 60). Many modifications of this arrangement have been described (Fig. 10, G).

Another special modification of the gut is met with in the larvae of Hymenoptera-Apocrita, and in the larvae of *Mymeleon* and other Neuroptera (Fig. 10, H). In these the mid-gut remains unconnected with the hind-gut until the time of pupation. Consequently, the stomach becomes enormously distended with the undigested residue of the food, which is not evacuated until just before pupation.

Proventriculus

In nearly all insects there are special structures where the fore-gut and the mid-gut join. There is nearly always a muscular sphincter by which the contents of the two regions can be kept separate; and there is often a discrete organ known as the proventriculus. The proventriculus is a different morphological entity in different groups, sometimes being composed solely of fore-gut, while sometimes both fore-gut and mid-gut contribute to it. Its function also varies. In the Blattidae it is a powerful muscular organ, or gizzard, provided with rows of massive teeth, which seize and triturate the solid particles of food in the crop. In the bee and other Hymenoptera the proventriculus is termed the 'honey stopper'. It forms an elongated plug with an elaborate musculature. The lumen bears rows of curved spines by means of which the pollen grains are filtered out from the nectar and passed on to the stomach while the clarified nectar is retained in the crop. In the flea it is lined with long needle-like teeth directed backwards; during digestion it contracts rhythmically, driving the teeth backwards into the mid-gut; at the same time the mid-gut contents are driven up against it by anti-peristaltic waves, and in this way the blood corpuscles are pounded and disintegrated.

In most insects the fore-gut is invaginated to some extent into the mid-gut, and this invagination, surrounded by the anterior or cardiac region of the mid-gut, is also commonly referred to as the 'proventriculus'. The invaginated fore-gut is called the 'oesophageal valve', though it seems never to function as a valve (if, by that word, is intended a mechanism which automatically allows movement in one direction only), for the occlusion of the passage between fore-gut and mid-gut is secured always by a sphincter muscle, and the invagination of the fore-gut has quite another significance.

Peritrophic Membrane

The gut of insects contains no mucous glands; the boluses and hard particles of food are therefore not lubricated as they are in the intestine of vertebrates; and the epithelial cells obtain protection in another way. As was first observed by Lyonet (1762), they are separated from the gut contents by a delicate membrane, composed of chitin mixed or combined with protein, known as the peritrophic membrane. It is this membrane which takes the place of mucus. Chito-protein is a mucopolysaccharide; mucus is a variable mixture of mucoprotein and mucopolysaccharides – so that chemically there is no great divergence here.

The peritrophic membrane is freely permeable to digestive enzymes and to the products of digestion; experimentally it has been shown to be permeable to all dyes save those with the largest colloidal particles. In its primitive state it is a somewhat indefinite affair formed by the condensation over the surface of the food of certain constituents in the mid-gut secretion; and it arises in this way to a greater or less extent in many existing insects, where it is made up of a number of concentric layers, all of which are chitinous. These membranes become separated from the general surface of the gut, but tend to remain attached at the anterior end, and are therefore apt to give a false impression of taking origin solely at this point. They remain attached in this region largely because here the mid-gut is overlaid by the oesophageal valve. That is probable the *raison d'etre* of this invagination: it enables the peritrophic membrane to extend forward beyond the point at which the food enters the mid-gut.

Now as the food passes through the oesophageal valve it will tend to press the walls outwards; and in so doing it will tend to press the secretion, from the cells of this cardiac region, into a membrane. That is the crudest form of mechanism by which the peritrophic membrane is pressed out; it is met with in the larvae of Lepidoptera. But in many insects the conditions have become far more elaborate. The cells in the cardiac region have become increasingly specialized for the production of the membrane until, in the Diptera and in the Dermaptera, they alone are concerned in its formation. At the same time the mechanisms by which the fluid secretion from these cells is pressed to form a solid tube become more complicated, until in many Diptera these annular moulds or presses present the most elegant forms. Sometimes the oesophageal valve bears a rigid cuticular ring against which the cells of the mid-gut are forced (Fig. 11, A, C); sometimes the valve is itself a solid structure made up of large vesicular cells (Fig. 11, B); sometimes, a rather cruder arrangement, the valve is thin-walled but contains blood sinuses which can be distended with fluid and so blown out against the cardiac cells.

The peritrophic membrane is present in the majority of insects. It

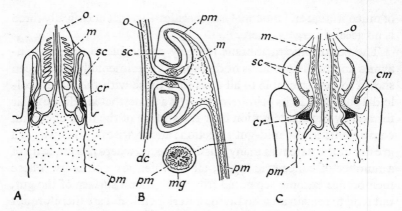

Figure 11. Annular moulds producing peritrophic membrane: A, larva of mosquito, *Anopheles*; B, tsetse-fly, *Glossina*; C, earwig, *Forficula*. (Modified after Wigglesworth.) The cross-section of the membrane is equal to that of the mould in each case; hence the complicated folding of the membrane in the narrow part of the mid-gut in B

cr, cuticular ring forming inner wall of press; *cm*, circular muscle compressing the outer wall against this ring; *dc*, duct of crop; *m*, sphincter muscle; *mg*, mid-gut; *o*, oesophagus; *pm*, peritrophic membrane; *sc*, cells secreting the substance of the membrane

has disappeared in most Hemiptera, perhaps because these take only liquid food, but it is present in some Corixids which feed on detritus and other solid matter; it is said to be absent in the blood-sucking adults of Tabanidae, fleas and sucking lice, and it is exceedingly tenuous and delicate in the adults of Culicidae, Simuliidae, *Phlebotomus*, &c. It is wanting also in the Carabidae and Dytiscidae, possibly because their digestion is largely extra-intestinal (see p. 53), possibly because their secretion is of the holocrine type (see p. 54) in which the secreting cells break down completely during the process. It is said to be absent also in adult ants – but here again, if it is very delicate, it may have been overlooked.

Salivary Glands

Most insects possess salivary glands. Sometimes, as in the Apterygota and in the bee, several kinds of glands, probably with very different functions, open in the region of the mouth; and even the single pair of labial glands present in most insects have very varied functions. (i) In the first place, their secretion may serve to moisten and dissolve the food: when a cockroach eats, its mouthparts are regularly moistened with saliva and in the feeding butterfly a drop of saliva is extruded from the tip of the tongue. (ii) Their secretion often contains digestive enzymes which may act outside the body and which continue to act upon the food after it has been swallowed, as in the case of amylase in the cockroach or invertase in butterflies and bees. The saliva of plant-sucking bugs commonly contains invertase and amylase, as in most Aphids; but in some Capsids and Jassids it may contain also lipase and protease which cause much necrotic injury to the plants on which they feed. (iii) In the Hemiptera, Diptera and Siphonaptera, which have mouthparts adapted for sucking up fluids, the saliva is discharged at the tip of the proboscis, and one of its functions may be to keep the tube, up which the fluids are sucked, moist and clean. (iv) And in those forms which suck the blood of living animals, the saliva usually contains an anticoagulin – probably antithrombin. It has been shown that if the salivary glands are removed from the living tsetse-fly, although it continues to take food for some time, ultimately its crop and proboscis become blocked with coagulated blood. (v) Finally, in many insects the labial glands are not concerned with feeding and digestion but become modified for the production of silk or for other purposes.

Stylet sheath. The plant-sucking bugs *Oncopeltus* and *Dysdercus* produce two types of saliva coming from different lobes of the salivary glands: a watery saliva containing the digestive enzymes, and a viscid material which lines the path made by the stylets through the tissues of the plant to form a 'stylet sheath'. This appears to be composed of self-tanned material resembling the cuticulin of the epicuticle (p. 5). It serves to filter off particles in the sap which might obstruct the sucking canal and to prevent the sap from exuding to the exterior around the stylets.

Extra-intestinal digestion. It is interesting to note that many predaceous beetles (larvae of *Dytiscus*, *Carabus* and *Cicindela*, and the adults of these forms) which lack salivary glands, as well as the ectoparasitic larvae of certain Hymenoptera, eject their intestinal secretion from the mouth, allow much of the digestion of their prey to take place outside the body, and then reabsorb the fluid products. This process is termed extra-intestinal digestion. The flesh-eating maggots achieve the same result by discharging proteolytic enzymes in their excrement. The assassin bug (*Platymeris*) discharges a toxic and strongly lipolytic and proteolytic saliva which quickly paralyses the prey and liquefies its tissues.

Some insects, when they emerge from the pupa, discharge secretions which soften the cocoon and facilitate the escape of the adult. That has long been known in the silkworm. In *Antheraea* the moth produces a solid protein (consisting of almost pure protease) secreted from the maxillary galeae, which is dissolved by a copious alkaline secretion from the labial glands. In many gall-producing insects the salivary secretion contains indolacetic acid which evokes abnormal growth of the plant tissues. In robber flies, Asilidae, the salivary glands secrete a neurotoxic venom resembling cobra venom.

Secretion

The secretion of digestive juices in insects is commonly stated to be of two histological types: holocrine and merocrine. In the former type, which occurs notably in Orthoptera and Coleoptera, the active epithelial cells disintegrate completely in the process of secretion, and are replaced by the growth of new cells from special cell nests or nidi. In the merocrine type, the cells do not break down. They may discharge their secretion, usually through a striated border, without any visible changes in the cell (save what may be detected by refined cytological methods) and that is probably the most common state of affairs. But a special type of merocrine secretion, known as vesicular secretion, is described, in which discrete droplets of fluid are eliminated through the striated border. Unfortunately, identical droplets are only too easily produced artificially during fixation, and appear in abundance where during life they cannot be seen at all. But this does not, of course, exclude the possibility that in some cases vesicular secretion may actually take place. Sometimes the cells of the gut

contain conspicuous vacuoles, and these are often taken as evidence
of secretory activity. But it must not be forgotten that such vacuoles
may equally well arise in the course of absorption. In Aphides, fed
after a period of fasting, successive waves of secretory activity appear
to pass down the stomach.

Reaction of the Gut

As was discovered long ago by Plateau (1874), the intestinal contents
of insects are not as a rule strongly acid or alkaline. The saliva in
those forms in which it has been examined is usually neutral; the gut
contents usually slightly acid, having a pH between 6 and 7. The
contents of the gut tend, on the whole, to be more alkaline in phytophagous insects (pH 8·4–9·0 in some herbivorous Coleoptera, pH
8·4–10·3 in the silkworm) and more acid in carnivores; but there are
many exceptions. Well-marked acidity (pH 4·8–5·2) may develop
in the crop of the cockroach, after a meal of carbohydrates, as the
result of acid fermentation by micro-organisms; and sometimes the
insect itself is responsible for a quite strongly acid secretion; as in
a well-defined region in the intestine of the blowfly maggot, which
has a pH of 3·0–3·5, the acidity being due probably to phosphoric
acid.

Digestive Enzymes

The digestive enzymes of insects are, in general, such as might be
predicted from the nature of their food. Omnivorous forms, like the
cockroach, have a full complement of enzymes capable of digesting
all the common foodstuffs. Insects which take chiefly a protein diet,
such as the exclusively blood-sucking species, have little but proteolytic enzymes. Where an insect feeds only on nectar, as in the adults
of Lepidoptera, only an invertase is present; whereas the phytophagous larvae of these same forms have a protease, lipase, amylase,
maltase, and invertase. Table I gives some examples in simplified
form to illustrate this relationship, the enzymes being given the
names that were used in the past. The cockroaches *Blaberus* and

Periplaneta are described in current terms as producing α-glucosidase (acting on sucrose and maltose), β-glucosidase (cellobiose), α-fructofuranase (raffinose and sucrose), α-galactosidase (melibiose), and β-galactosidase (lactose). Many insects secrete a chitinase into the mid-gut; this enzyme seems to be identical with the 'lysozyme' which dissolves the cell walls of micro-organisms and is responsible for the bacteriolytic activity of the gut contents.

These enzymes are readily demonstrated in extracts from the gut. They are present also in the juice removed from the lumen, showing that at least the greater part of digestion is extracellular. As can be seen from Table I most of the enzymes are secreted in the mid-gut; the salivary glands usually secrete only amylase, and often no enzymes at all; but we have already referred to the varied enzymes in the salivary glands of some plant-sucking Hemiptera (p. 53).

The properties of the enzymes of insects are, in general, very like those of the corresponding enzymes from vertebrates. They are similarly affected by changes in hydrogen ion concentration, and are stimulated or inhibited by the same factors. The chief proteolytic enzymes seem almost always to be more or less like the trypsin and chymotrypsin of vertebrates though they often act best over a rather more acid range of pH. As in vertebrates the peptidases are of three sorts, 'aminopolypeptidase' and 'carboxypolypeptidase' characterized by their ability to break off amino-acids from the amino end or the carboxyl end of the polypeptide chain, and 'dipeptidase' which hydrolyses dipeptides. Enzymes of the pepsin type acting in a very acid medium (pH 2–3) are found in the maggots of *Calliphora* and other flies. Certain special enzymes are present in those insects which feed on particular substances; for instance, the meat-eating maggot (*Lucilia*) has a collagenase which attacks raw connective tissue in alkaline medium; and a few of the wood-boring beetles produce a cellulase. The clothes moth (*Tineola*) can digest keratin; it contains in its gut a powerful reducing system which will effectively break the sulphur linkages by means of which the adjacent polypeptide chains in keratin are bound together. Once reduced in this way the chains are hydrolysed by means of a protease of tryptic type which is itself resistant to the very low oxidation-reduction potential

Table 1

| Insect | Food | Enzymes | |
		Salivary glands	Mid-gut
ORTHOPTERA			
Cockroach (*Periplaneta americana*)	Omnivorous	Amylase	Amylase: maltase; invertase; tryptase; peptidase; lipase
COLEOPTERA			
Japanese beetle (*Popillia japonica*)	Foliage	—	Maltase; invertase: tryptase; lipase
HYMENOPTERA			
Honey-bee (*Apis mellifica*)	Nectar and pollen	Amylase; invertase	Amylase; invertase; tryptase; lipase
LEPIDOPTERA			
Silkworm (*Bombyx mori*) Larva	Foliage	—	Amylase; maltase; invertase (chiefly in the cells); tryptase; peptidase (only in the cells); lipase (in very small amounts)
Humming bird hawk moth (*Macroglossa stellatarum*) Adult	Nectar	Invertase	Invertase
DIPTERA			
Blowfly (*Lucilia sericata*) Larva	Flesh	Amylase (very weak)	Tryptase; collagenase; peptidase; lipase
Blowfly (*Calliphora*) Adult	Liquid food of all kinds	Amylase	Amylase; maltase; invertase; weak tryptase and peptidase
Chrysops silacea Adult	Female: blood Male: nectar, &c.	Nil	Very weak amylase; strong invertase; tryptase; peptidase
Tsetse-fly (*Glossina*)	Blood	Nil	Very weak amylase; strong tryptase; peptidase

in the gut. The bee moth (*Galleria*) can digest all the components of beeswax (except paraffin), but certain bacteria in the gut are largely responsible for this.

Digestion of cellulose. So many insects feed on the foliage and woody tissues of plants that particular interest attaches to the question of cellulose digestion. In the vast majority of species, the cellulose is quite unaffected by passage through the intestinal canal: those plant cells that are not ruptured by the jaws of the caterpillar pass through the gut with their walls intact, although their contents may be digested; the excrement of many wood-boring beetles contains all the lignin and cellulose that have been ingested with the food; the very starch grains may be protected from digestion by their pectin covering. On the other hand, many insects can digest cellulose; sometimes by means of cellulase they secrete themselves, sometimes with the help of symbiotic micro-organisms; and hemicellulose (lichenin, &c.) is digested by most phytophagous insects.

Symbionts in Digestion

Many insects have a rich fauna of bacteria or protozoa in the gut, but in none is this so prominent as in the hind-gut of wood-boring termites. Here there is an amazing population of flagellates, ciliates, and spirochaetes, which perform for their host the invaluable function of digesting cellulose. Termites with this fauna intact can thrive indefinitely upon pure cellulose; but if the insect is deprived of these organisms (by exposing it to a high tension of oxygen, for example) its powers of digesting cellulose are completely lost, and unless re-infected it soon dies. Cellulase can be extracted from the flagellates in the gut of these termites and of certain cockroaches (*Cryptocercus*); whereas this enzyme is absent from defaunated insects.

Certain Lamellicorn beetle larvae (*Cetonia, Oryctes, Osmoderma*) that feed on pine-needles and such like, ingest with their food those micro-organisms which ordinarily ferment cellulose in nature. These florish in the dilated hind intestine (the so-called 'fermentation chamber') of these larvae, and form an important, perhaps an essential, aid to digestion – being themselves digested later by the

proteolytic enzymes of their host. The same occurs in some Tipulid
larvae. But there are related beetles (*Dorcus*) which have a similar
fauna and yet are said not to digest cellulose. Other insects that feed
on wood harbour yeast-like organisms, often within the cells of the
gut; and it was natural to attribute to these, also, the function of
digesting cellulose. But, in fact, some beetles that have such 'symbionts' are unable to digest cellulose, and others which are without
them can do so. The food fungus cultivated by the leaf-cutting ant
Atta in its fungus gardens likewise serves to break down cellulose.

Many blood-sucking insects, also, possess intra-cellular bacteria
or yeasts in their tissues; and the occurrence of these in close
association with the gut, in the case of the tsetse-fly (*Glossina*) and
the Pupipara, led to the suggestion that they were concerned in the
digestion of blood. On the face of it this was not a probable hypothesis, for the blood proteins do not differ specially from those of
other tissues; and it has been shown (at least in the tsetse-fly) to be
incorrect; for no digestion of the blood takes place in that anterior
part of the mid-gut in which the symbionts occur. The function of
these symbiotic organisms is to be sought elsewhere (p. 81).

Absorption

In the cockroach, in which the food is digested largely in the crop, a
small amount of absorption, notably of fats, may also occur in that
segment of the fore-gut; but the greater part of absorption undoubtedly takes place in the mid-gut. The cells seem never to take up solid
particles; the dissolved foodstuffs, like the digestive enzymes, diffuse
through the peritrophic membrane, when such is present, and are
absorbed by the epithelial cells. Absorption in the mid-gut of the
cockroach or the locust seems to be largely a matter of 'facilitated
diffusion'; for example, as soon as glucose or other sugars enter the
circulating blood they are converted into trehalose. A steep gradient
of concentration of the sugar in question as between the gut contents
and the haemolymph is thus maintained and facilitates uptake by
diffusion.

Hindgut absorption. In the blood-sucking insects, such as the
tsetse-fly or the mosquito, almost nothing is passed on to the hind-

gut, save a little haematin. In these forms the hind-gut clearly plays no part in absorbing foodstuffs; but whether this is generally true is uncertain. We have seen that the greater part of digestion in wood-eating termites and Lamellicorns takes place in the hind-gut, to what extent the products are absorbed here, or returned to the mid-gut, is not known; but the hind-gut of Lamellicorn larvae has special areas of cells which are believed to be concerned with absorption.

Water is certainly absorbed in the hind-gut and rectum of many insects (p. 71); and the food residue, in contact with the rectal glands, may be converted to a more or less dry faecal pellet.

Besides water inorganic ions are actively absorbed by the rectum and this function is important in the regulation of osmotic pressure and ionic composition in the blood (p. 76). In addition, amino acids can be absorbed in the rectum of the locust *Schistocerca*, but whether other substances besides water and inorganic ions are absorbed in the hind-gut of most insects is not known.

The Faeces

The excrement of insects varies in character from dry pellets to a copious clear liquid, according to the amount of fluid in the diet. In plant-sucking Aphids and Coccids it contains much unabsorbed organic matter, particularly carbohydrates and amino acids; the dried residue collects on the surface of plants to form 'honey-dew', manna and such-like products. These insects take in phloem juice which is under pressure in the plant, absorb perhaps no more than half the amino acids and amides in the sap and even less of the sugar, and excrete the residue containing a wide range of carbohydrates that have been synthesized in the gut by the 'transglucosidase' activity of invertase acting on the sucrose of the plant sap.

BIBLIOGRAPHY

HOUSE, H. L. *Physiology of Insecta*, II (Morris Rockstein, Ed.), Academic Press, New York, 1964, 815–858 (insect digestion)
LIPKE, H. and FRAENKEL, G. *Ann. Rev. Entom.*, **1**, (1956), 17–44 (insect nutrition: review)

TREHERNE, J. E. *Viewpoints in Biology* (J. D. Carthy and C. L. Duddington, Ed.), **1**, (1962), 201–241 (absorption from the gut of insects: review)

WATERHOUSE, D. F. *Ann. Rev. Entom.*, **2**, (1957), 1–18 (digestion in insects: review)

WIGGLESWORTH, V. B. *Tijdschrift Ent.*, **95**, (1952) 63–68 (symbiosis in blood-sucking insects: review)

—— *The Principles of Insect Physiology*, 7th Edn., 1972, Chapman and Hall, London, 476–552 (digestive system of insects)

5 Excretion

The function of the excretory organs of animals is to maintain a more or less constant 'internal environment'. To this end they may be called upon to eliminate several classes of substances: (i) substances, like mineral salts or water, which are present in simple excess in the diet; (ii) the end products in the metabolism of organic nitrogen, sulphur and phosphorus, of which, particularly of nitrogen, there is always a large surplus for excretion – these form perhaps the most important category of excretory substances; (iii) more or less complex compounds, not uncommonly pigmented, which arise perhaps as accidents or by-products in the course of other chemical changes, and which have to be excreted because the chemical equipment of the body is powerless to deal with them; (iv) and, finally, in order to maintain such properties of the blood as the hydrogen ion concentration, or the osmotic pressure, the excretory organs may be called upon to remove certain substances such as acids or bases, and to hold back or reabsorb others such as water or specific ions.

Excretion of Dyes

Organs may fairly be regarded as excretory if they perform these functions, or eliminate one or more of these classes of substances. But, in practice, it is not always easy to recognize these activities; and therefore a second, experimental, criterion of excretory organs has come to be used – the excretion of dyes introduced into the body. This test must be used with discretion; for different dyes are absorbed by different tissues; methylene blue, for example, is taken up by the nervous system, by the tracheal cells and by the oenocytes, and is therefore of no use for this purpose. The most reliable dye is indigocarmine, which seems to be removed specifically by the excretory organs of all animals. When indigocarmine is injected into insects, it rapidly appears in the Malpighian tubes, which are unquestionably excretory organs. In the primitive Apterygota it appears, also, in the lower segment or labyrinth of the labial glands, and it is probably correct to regard these also as excretory organs, homologous perhaps with the excretory glands of Crustacea. That is all the information obtainable with indigocarmine.

But if the acid dye ammonia carmine (which, in vertebrates, is filtered through the glomerulus of the renal tubule) is injected into insects, it is not excreted by the Malpighian tubes, but is absorbed and segregated by special cells – notably, by the 'pericardial cells' which lie along the dorsal heart and aliform muscles, by the 'garland-like strand' discovered by Weismann (1864) in the Muscid larvae, by various scattered groups of cells, and, in the Apterygota, by the upper segment or saccule of the labial gland. On this evidence Kowalevsky (1889) put forward the view that these cells are excretory organs (so-called 'acid excretory organs', because litmus is turned red within them) analogous in function to the kidney glomerulus.

'Nephrocytes'

Kowalevsky's experiments were so striking that this theory at once gained acceptance; but it is doubtful if it will bear close examination. Our ideas on the kidney glomerulus have changed so much in the last eighty years that Kowalevsky's analogy carries no weight at all. If the ammonia carmine cells are really 'storage kidneys' (nephrocytes, athrocytes) as the theory implies, the waste substance which they accumulate during life should at least be visible. But in many insects nothing seems to accumulate within them at all. Often they contain pigmented granules; but this is the case with many cells – in the fat body, epidermis, tracheal matrix and so forth. Occasionally they may become laden with some pigment from the food; in *Rhod-*

nius they are filled with blue-green granules of biliverdin derived from the breakdown of traces of haemoglobin that have been absorbed unchanged from the gut; and it cannot be denied that they are then serving as 'storage kidneys' for these particular substances; but this can scarcely be their chief function.

Functions of pericardial cells. Perhaps they play some undetermined part in intermediary metabolism; or they may be excretory organs in the sense of synthesizing the waste products that are to be eliminated by the Malpighian tubes. They are highly capricious in their absorptive properties: they will take up various proteins; they will absorb some dyes, such as trypan blue and ammonia carmine, but not others; and it has not been possible to recognize any common feature in molecular structure or physical or electrical properties to account for these differences. These properties are doubtful evidence of an excretory function. In vertebrates these same properties are shown by cells of the reticulo-endothelial system, and the pericardial cells of insects are commonly regarded as being equivalent to these. They seem to be a major site for the turnover of unwanted proteins. We have noted already their function of producing a cardiac accelerator in the cockroach (p. 37).

Urate Cells and Storage Excretion

As was observed by Milne Edwards and later by Fabre (1863), there are cells in the insect body which become laden with crystalline spheres of uric acid ('urate cells'); and they surely can be regarded as storage kidneys, for uric acid is an undoubted excretory substance. In many cases this view is probably correct; throughout larval life in the social Hymenoptera, uric acid collects in solid form in special urate cells scattered throughout the fat body, and is not transferred to the Malpighian tubes (of the adult) until the end of pupal life; in *Lepisma*, Dermaptera and many Orthoptera, this state of affairs persists throughout life, and the Malpighian tubes discharge very little uric acid.

Even here the observations must be interpreted with caution. For uric acid is the end product of protein katabolism; and when protein

breakdown is occurring very actively within a cell, it is likely that the
rate of formation of uric acid may, on occasion, exceed the rapidity
with which it can diffuse from the cell and be carried away by the
blood; particularly if conditions in the cell become unduly acid.
The uric acid will then crystallize out, and its subsequent solution
and removal will take place very slowly. That indeed is what appears
to happen in the cells of the fat body in Muscidae during metamorphosis: granules of uric acid ('pseudonuclei') crystallize out
within the 'albuminoid' deposits, and not until late in pupal life is
the uric acid trasferred to the Malpighian tubes. The same thing
happens in Lepidoptera; and here the uric acid must be endogenous,
for it is not increased by the artificial introduction of preformed uric
acid. In the larva of the mosquito *Aëdes* the cells of the fat body
become laden with uric acid in the fasting insect; but as soon as the
larva is abundantly fed the cells of the fat body are filled with
reserve food substances and the uric acid is displaced.

It is probable that not a few of the urate cells of insects are of this
nature – not true excretory organs but active cells in which uric
acid has crystallized out, as it were, by accident. For instance, the
epidermal cells of some Hemiptera contain uric acid; but it has been
shown that in one case at least, that of *Rhodnius prolixus*, the deposition of this uric acid takes place only at one stage of the moulting
cycle – just before the new cuticle is laid down; that is, when the cells
in question are most active in producing chitin – perhaps from the
proteins of the food.

Pigments as Excretory Products

Besides this uric acid, there is in the epidermal cells of *Rhodnius*
a red pteridine pigment, which appears at the same time as the uric
acid, and then slowly diminishes. Perhaps this substance, also, is a
by-product of the synthetic activity of these cells; and perhaps that
is the origin of many of the pigments of insects – accidents in metabolism, which only secondarily acquire their biological significance.
Such substances as the pteridine pigments of many types that occur
in the wing scales of Pieridae and other Lepidoptera, or beneath the
integument of wasps, the anthraquinone derivatives which form the

red colour of the cochineal insect, and the remarkable pigments of
Aphids (aphins), also complex derivatives of anthracene, are probably of this type. But the pteridines in the wing scales of Pierid
butterflies contain as much as 14 per cent of the waste nitrogen
eliminated (mainly as purines) during the pupal stage.

The red and brown eye colours of insects (ommochromes) are
derivatives of kynurenine, itself an oxidation product of the
important amino-acid tryptophane. The green pigments of insects
(insectoverdins) are usually mixtures of the blue biliverdin (presumably derived from the breakdown of haemoglobin or cytochrome) and some yellow carotinoid obtained from the food. Some
insect pigments, such as the flavones in the wings of Satyrine
butterflies, are wholly derived from preformed substances already
present in their food plants.

Verson's Glands

It has been observed that when the silkworm casts its skin, its body
may be powdered with crystals of oxalate and uric acid derived from
the moulting fluid; and this led to the suggestion that, at the time of
moulting, the Malpighian tubes are thrown temporarily out of
action, and their excretory function taken over by the dermal
glands (Verson's glands, p. 5). But this idea has been shown to be
mistaken; the excretory substances in the moulting fluid really come
from the Malpighian tubes, escape through the anus beneath the
old cuticle, and so spread over the surface of the body. (It is just
possible that the calcareous warts of Stratiomyid larvae (p. 3) may
also be formed in this way. On the other hand, the lime which
appears in the ecdysial fluid of certain Diptera at the time of
pupation is said to be dissolved from the Malpighian tubes, reabsorbed into the blood, and excreted through the newly-formed
cuticle.)

Malpighian Tubules and the Urine

When all these subsidiary processes have been considered, the Malpighian tubules still remain unquestionably the chief excretory

organs. The Malpighian tubules are relatively simple tubular glands which open at the junction between the mid-gut and the hind-gut. They are exceedingly variable in form: sometimes being numerous (e.g. over 100) and short, sometimes few in number (e.g. 2) and long; sometimes simple and sometimes branched; occasionally anastomosing to form closed loops; while sometimes more than one type of tubule is present.

Histology. Their histological structure is no less variable. Usually their epithelial cells bear a striated border, but this may be wanting; sometimes this border is of the type known as a 'honey-comb border', or 'Wabensaum', in which the rod-like filaments or 'microvilli' are so closely packed, being separated by no more than 15–20 nm, that they appear with the light microscope as though fused together; sometimes the microvilli are more widely separated and the mobile filaments form a 'brush border' or 'Burstensaum'. The cytoplasm of the Malpighian tubules is exceedingly rich in mitochondria and these commonly extend far into the microvilli, so that it sometimes appears as though the striated border were composed of mitochondria.

Often the tubules have a muscular coat and are capable of active movements; sometimes they show only slow twisting movements due to changes in the secretory pressure within them. Many histological changes have been described in the active cells of the Malpighian tubules: the discharge of vesicles, the eruption of vacuoles, and so forth. It is not improbable that more than one cytological mechanism of excretion may exist; but certainly many of the recorded observations are artefacts. Studies with the electron microscope have revealed the uptake of fluids from the lumen by a process of pinocytosis between the bases of the microvilli; and also the discharge of minute vesicles, apparently from the vesicular endoplasmic reticulum, at the tips of the microvilli. In *Drosophila* there is a regular flow of vesicles from the basement membrane to the lumen.

The urine. The fluid which the Malpighian tubules produce is equally varied. It may be a clear fluid or a thick pasty suspension:

the colourless fluid passed by blood-sucking insects soon after a meal is an example of the former sort, the 'meconium' of newly emerged Lepidoptera of the latter. Its constituents depend naturally upon the nature of the food, but it always consists primarily of water with the usual inorganic salts, chlorides and phosphates of sodium, potassium, calcium, and magnesium, in solution, or in suspension. Yellowish or greenish pigments ('entomo-urochrome') are often present; but it is likely that these differ chemically in different species: riboflavine is plentiful in the cells of the Malpighian tubules of some insects, kynurenine in others.

Nitrogen excretion. Nitrogen is excreted chiefly as uric acid; but urea has been reported in substantial amounts in the urine of the clothes moth, and meat-eating maggots eliminate much of their nitrogen as ammonia. Some aquatic insects, such as the larvae of *Sialis*, of Odonata and Trichoptera also eliminate their nitrogen in the form of ammonia; and so indeed do Aphids, which are continually absorbing large quantities of water from the plant and discharging copious amounts of 'honey dew'. Guanine, the main nitrogenous constituent in the urine of many Arthropods, seems not to occur in insects; but many other nitrogenous compounds such as xanthine and hypoxanthine may occur; and many insects will break down their uric acid, wholly or in part, to allantoine or allantoic acid. Allantoine is plentiful in the excreta of blowfly maggots. The enzyme uricase which converts uric acid to allantoine is particularly active in many Heteroptera (e.g. *Dysdercus*) which excrete most of their nitrogen in this form. Aphids and the tsetse fly *Glossina* eliminate substantial amounts of nitrogen in the form of the nitrogen-rich amino acids arginine and histidine.

The reaction of the urine naturally varies with the diet, and the period after feeding; the thick meconium of Lepidoptera has a pH of 5·8–6·3; the similar urine of the bug *Rhodnius prolixus* shows the same kind of range.

When uric acid is present in the urine it is usually in the form of crystalline spheres with a radial striation. These spheres are commonly stated to consist of ammonium acid urate, or of sodium or potassium acid urate; but in *Rhodnius*, almost the only insect in which this question has been investigated chemically, they consist mainly of free uric acid; and it is possible that this is also the case in other insects.

Other granular contents. Besides these crystals of uric acid, the Malpighian tubules may contain solid spheres or granules of calcium or magnesium carbonate and phosphate and, notably in the larvae of Lepidoptera, crystals of calcium oxalate. The oxalate is possibly derived from the preformed oxalic acid in the food, but the significance of the accumulated carbonates is uncertain. They are sometimes used by the pupating larva to strengthen its cocoon (Cerambycidae, Col.) or its puparium (Agromyzidae, Dipt.) or by the female to calcify the shells of her eggs (Phasmidae, Orth.), but these are probably secondary adaptations. It has been suggested that the carbonates may constitute a method of eliminating carbon dioxide; but such a mechanism seems uncalled for, and in any case the amount of carbon dioxide bound in this way must form but a fraction of what the insect produces. Perhaps these carbonates simply provide a mechanism for getting rid of excess alkali; it is at least noteworthy that they occur chiefly in saprophagous and phytophagous larvae of Diptera (Stratiomyiidae, Drosophilidae, Agromyzidae, and many more) whose food might be expected to contain an excess of fixed base. Their urine must therefore be alkaline, and in the presence of the carbon dioxide of the tissues, if calcium or magnesium ions are present in quantity, carbonates will tend to precipitate. It is worth noting that carbonates are present in the phytophagous Agromyzid *Acidia*, but not in the parasitic genus *Cryptochaetum*; and they are said to be absent in aquatic larvae from very acid waters. (It may be noted in passing that similar deposits of lime ('calcospherites') may occur in the fat body of some insects. In Aphids the mid-gut cells become filled with inclusion bodies consisting or carbonates and phosphates of calcium and magnesium.)

Excretion in Rhodnius

Insects are essentially terrestrial animals. Consequently, the main problem with which their excretory system is faced is the elimination

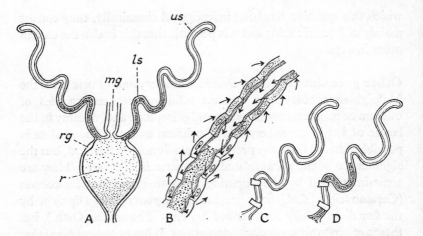

Figure 12. A, excretory system of *Rhodnius*: schematic; B, detail of Malpighian tubules at junction of upper and lower segments; arrows indicate the circulation of water and base; C, two ligatures applied to Malpighian tubule three hours after meal; D, the same 24 hours later. (Modified after Wigglesworth)

ls, lower segment of Malpighian tubule: lumen full of uric acid spheres; *mg,* mid-gut; *r,* rectum; *rg,* rectal gland; *us,* upper segment of tubule

of all these substances, and particularly the elimination of nitrogen, in such a way that they are able to conserve their limited supplies of water; and it is instructive to consider the excretory system from this point of view. It will be convenient to start by describing the excretory system of the blood-sucking bug *Rhodnius prolixus*, which has been studied in considerable detail, and then to compare with this some other insects.

The excretory system of *Rhodnius* consists of four very long Malpighian tubules, opening into a large rectal sac (Fig. 12, A). Each tubule is made up of two distinct segments: a translucent upper segment, about two-thirds of the whole, the lumen of which contains only a clear fluid, and an opaque white lower segment stuffed with spheres of uric acid. The histological structure of these two segments is quite different; the one changes abruptly into the other; and immediately below the junction the granules of uric acid appear in the lumen (Fig. 12, B). The explanation of these facts which has been suggested is that the upper segment of the Malpighian tubule is secreting a solution of sodium or potassium acid urate, and the lower segment reabsorbing both the water and a large proportion of the base (sodium or potassium – perhaps in the form of bicarbonate) leading to a precipitation of the free uric acid. For it can be shown (i) that the contents of the upper segment ($pH = 7\cdot2$) are definitely more alkaline than those of the lower segment ($pH = 6\cdot6$); (ii) that neutral red added to the blood of the insect is secreted into the lumen by the cells of the upper segment and reabsorbed again by the cells of the lower segment; and (iii) that if two ligatures are applied to the tubule near its lower end at a time when the uric acid spheres have been washed out by a recent meal, then uric acid appears above the upper ligature and below[1] the lower ligature, but between the ligatures there is no uric acid, nor is there any distension of the tubules (Fig. 12, C, D).

Thus it is suggested that, in *Rhodnius*, there is a continuous circulation through the excretory system of both water and of base, the same water and base being used over and over again to carry uric acid from the body.

It is not unlikely that the rectum, also, assists in the process of reabsorbing water; for around the point of entry of the Malpighian tubules there is a ring of large epithelial cells constituting a so-called rectal gland; and we shall find that there is evidence in other insects that the function of the rectal glands is the absorption of water and ions from the excrement. Simultaneous measurements of the concentration of ions in the haemolymph and in the lumen of different parts of the excretory system in *Rhodnius* have shown that there is in fact a selective excretion of potassium into the lumen of the upper segment of the tubule and a reabsorption of potassium from the lower segment; water and sodium, but not potassium, being reabsorbed in the rectum during the later stages of excretion. In all insects potassium seems to play a prime role in activating or stimulating the secretory process in the Malpighian tubules.

Rhodnius takes in very large meals of blood (p. 98) and like most blood-sucking insects immediately proceeds to excrete the greater

---

[1] This has doubtless been derived from the other Malpighian tubes and entered from below.

part of the water and salts. This intense 'diuresis' is brought about by a hormone produced by neurosecretory cells in the brain and, particularly, in the fused ganglia of the thorax. The mere injection of saline into the body cavity of *Rhodnius* does not induce a rapid flow of urine in this way.

Excretion in Other Insects

This conception of a circulation of water through the excretory system is a familiar one in the physiology of birds, and reptiles, and mammals, where the renal tubule itself, the cloaca, and the large bowel all take part in the reabsorption of water. This idea, introduced into the physiology of insects, helps to explain many of the properties of their excretory system. Diuretic hormones (and sometimes anti-diuretic hormones) are produced also by neurosecretory cells in many other insects.

Rectal reabsorption. The simplest system that occurs is that shown in Fig. 13, A. Here the Malpighian tubules are composed of a single type of cell, and contain only fluid. As this fluid, mixed with the intestinal contents, passes down the hind-gut, water is absorbed from it – first by the cubical epithelium of the intestine and finally by the epithelium of the rectum. The material is retained in the rectum until it has been converted into a more or less dry pellet. In some cases, for instance in the human louse *Pediculus*, the contents of the mid-gut may be retained in the stomach so that the hind-gut contains only urine; under these circumstances, a little pellet of solid uric acid may be formed in the region of the rectal glands, clearly demonstrating the absorption of water by these organs. This arrangement is found in *Lepisma*, Dermaptera, Orthoptera, Neuroptera, and many beetles. In some cases the active epithelial cells of the rectum are collected into compact areas, the 'rectal glands'; in other cases they are evenly spread over the gut wall.

In the rectum of *Schistocerca*, *Periplaneta*, *Calliphora* etc. there is an active absorption of ions, notably potassium and sodium, from the rectum along with amino acids and water. The physiological mechanism of such absorption is under active investigation.

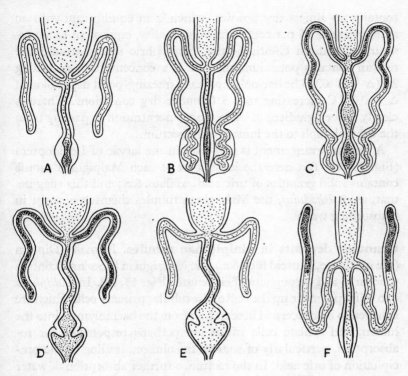

Figure 13. Types of excretory system in insects: schematic
Hind-gut indicated by a heavy line internally. Explanation in text

Cryptonephridial system. In many beetles, phytophagous, carnivorous, and omnivorous species, as well as those that feed on dry substances, such as the mealworm (*Tenebrio*), the upper parts of the Malpighian tubules closely invest the rectum, being bound to it by a delicate membrane (Fig. 13, B). The precise physiology of this arrangement is not fully known; but it is clear that it serves to add the absorptive powers of the Malpighian tubules to those of the rectal epithelium. The arrangement seems always to be associated with remarkable powers of drying the excrement. The Malpighian tubules of the mealworm produce a clear fluid but the rectum contains an almost dry powder (which is in equilibrium with an atmosphere of 75 per cent. relative humidity, equivalent to super-saturated sodium chloride); the 'perinephric space' around the rectum contains potassium chloride at a concentration exceeding $2M$ ($\Delta = 8°$ C). The haemolymph has a freezing-point depression of $\Delta = 0.75°C$ increasing to $1.5°C$ under dry conditions. There is clearly a steep gradient of increasing concentration in passing from the haemolymph to the lumen of the rectum.

A similar arrangement is met with in the larvae of Lepidoptera (Fig. 13, C). But here the lower part of each Malpighian tubule contains solid granules of uric acid, oxalate, &c.; and this suggests that, as in *Rhodnius*, the Malpighian tubules themselves assist in reabsorbing water.

Granular deposits in Malpighian tubules. In many Diptera (the mosquito, Muscid flies, &c.), the Malpighian tubes may contain solid uric acid throughout their length (Fig. 13, D). In *Calliphora* two cell types make up the wall of the tubule: primary cells which are believed to be concerned in secreting from the haemolymph into the lumen, and stellate cells which are perhaps responsible for re-absorption, particularly of sodium in solution, leading to the precipitation of uric acid. In the rectum, a further absorption of water and ions occurs; but these insects lack the power of really drying the excrement. Conditions are somewhat similar in certain Hymenoptera. In other Hymenoptera, the excretory system is like that of the flea (Fig. 13, E). It never contains any solid uric acid because sufficient water is retained in the capacious rectum to keep it in solution; but it is none the less probable that this fluid and also salts are reabsorbed, and circulate again through the system.

Finally, in those many dipterous larvae in which the Malpighian tubules contain carbonates, these are generally confined to the upper part of the tubule (Fig. 13, F); but the mechanism of their secretion is unknown.

These are just a few of the arrangements of the Malpighian tubes and hind-gut, which will serve to illustrate the general principles of excretion in insects so far as these are known at present; many other arrangements exist.

Excretion in Aquatic Insects

When insects come to live in fresh water they are no longer compelled to be so careful of their water supply. Water is taken up in quantity with the food, or it may enter through permeable regions of the body surface, such as the 'anal papillae' of mosquito larvae; and the Malpighian tubules produce a copious clear urine. In *Ilyocoris* and other aquatic Hemiptera the excretory system must work continuously to eliminate water gained by osmotic uptake through the cuticle.

In the mosquito larva this abundant supply of water does not appear to have resulted in any notable change in nitrogen metabolism: the fat body may at times become laden with uric acid, and if the uptake of water is restricted, as when the larva is kept in slightly hypertonic salt water, solid deposits of uric acid may appear in the lumen of the Malpighian tubules. On the other hand, certain aquatic insects, such as *Sialis*, Trichoptera, Odonata, *Dytiscus*, and *Notonecta* have become 'ammonotelic' in their larval stages and eliminate almost all their nitrogen as ammonia. They revert to a 'uricotelic' condition (uric acid excretion) during the pupal stage in preparation for terrestrial life.

Regulation of Osmotic Pressure and Ionic Composition of the Haemolymph

Life in fresh water raises new problems in the physiology of excretion. The continuous excretion of water is liable to deprive the insect of essential ions such as chloride, potassium, and sodium. Some aquatic larvae, such as *Sialis*, have a body covering which is highly impermeable to salts and there is presumably a very active recovery of salts in the rectum before the urine is discharged. Other larvae, such as that of the mosquito *Aëdes*, not only reabsorb chloride, sodium and potassium through the columnar epithelium lining the hind-gut, but they also have 'anal papillae' which can absorb the traces of chloride, sodium and potassium found in fresh water and so maintain actively the normal ionic composition of the blood. The plasma membrane of the papillae is deeply infolded and associated with mitochondria to form so-called 'mitochondrial pumps' which are concerned in the active and selective absorption of ions. The 'gills' of many other aquatic insect larvae have similar functions. In larvae of *Drosophila* and other Diptera there are flattened patches of enlarged cells close to the anus, which serve the same purpose.

In terrestrial insects (*Tenebrio, Schistocerca, Calliphora*, &c.) an active reabsorption of water and of inorganic ions by the rectal glands (p. 72) fulfils the same need to keep constant the osmotic pressure and the ionic composition of the haemolymph.

In the caterpillars of Lepidoptera the mid-gut plays an active part in ionic regulation. There is a potassium pump, perhaps located in the so-called 'goblet cells', which transport potassium from the haemolymph to the gut lumen. The mid-gut of *Periplaneta* has a comparable function.

Accessory Functions of the Malpighian tubules

Réaumur observed that the lemon yellow powder used by the larva of *Malacosoma neustria* to cover its cocoon is a product of the Malpighian tubules. Such products are used by other Lepidoptera: *Leucoma salicis, Eriogaster lanestris*, and so forth; and we have already seen how Cerambycidae strengthen their cocoons with lime from this same source. The foam or cuckoo-spit produced by the larvae of Cercopidae is formed out of a mucoprotein secretion from the Malpighian tubules. And in the larvae of such widely separated insects as the beetles *Phytonomus* (Curculionidae), *Lebia* (Carabidae), &c., and the larvae of Neuroptera (*Chrysopa, Sisyra*, &c.) a part of the Malpighian tubule is transformed into a silk gland used, among other purposes, for spinning the cocoon.

BIBLIOGRAPHY

CRAIG, R. *Ann. Rev. Entom.*, **5**, (1960), 53–68 (excretion in insects: review)

CROMARTIE, R. I. T. *Ann. Rev. Entom.*, **4**, (1959) 59–76 (insect pigments: review)

MADDRELL, S. H. P. *Adv. Insect Physiol.*, **8,** (1971), 200–331 (insect excretion: review)

SHAW, J. and STOBBART, R. H. *Physiology of Insecta*, III (Morris Rockstein, Ed.) Academic Press, New York, 1964, 190–258 (excretion: salt and water balance)

WIGGLESWORTH, V. B. *The Principles of Insect Physiology* 7th Edn. Chapman and Hall, London, 1972, 553–592 (excretion)

6   Nutrition and Metabolism

Since the alimentary canal of insects contains digestive enzymes
competent to hydrolyse those constituents commonly present in
their food (p. 55), it is probable that this food is absorbed in the
form of simple products of hydrolysis: carbohydrates as monosaccharides, fats as glycerol and fatty acids, protein as amino acids. But
at present there is no proof of this, and it is possible that there
may be some absorption of more complex bodies. In certain blood-
sucking insects, such as *Rhodnius*, although the greater part of
the haemoglobin is broken down to haematin in the lumen of the
gut, and the haematin residue eliminated unchanged, a small
amount of undigested haemoglobin is absorbed into the blood
and is subsequently broken down to bile pigments and other
products (p. 64).

The simple products of digestion provide the raw material for
growth and energy production. The sites of the intermediary stages
of metabolism and synthesis are not altogether known. It is certain
that the fat body plays a major part in the synthesis of trehalose,
the formation and storage of glycogen and fat, the deamination and
transamination of amino acids, the synthesis of protein and the
production of uric acid and other waste products (p. 44). But it is
unlikely that the fat body cells have a monopoly of these activities.
The epithelium of the gut wall can be the site of massive storage of
fats, glycogen, and protein. The haemocytes (p. 42) and pericardial
cells (p. 43) may well play a role in intermediary metabolism. And
such tissues as the epidermis and the growing muscles doubtless
take up the raw materials they require from the haemolymph and

accomplish a large part of the syntheses needed for their own activities; as already pointed out, they may well play a part also in deamination and the production of nitrogenous wastes (p. 65). Certainly the epidermis is concerned in the many complex synthetic activities necessary for cuticle formation.

Food Requirements

We have seen how exceedingly varied are the apparent food materials of insects (p. 46). But it must be remembered that insects are not always feeding upon what they eat: they may devour rotten wood but feed only upon the fungi which it harbours; or eat large quantities of cellulose and yet assimilate only the associated substances. The principles of insect nutrition are therefore not easy to state.

For growth and reproduction insects must have an adequate supply of carbon and nitrogen, phosphorus and sulphur, inorganic salts, and all the other elements that go to make up the living system. Aphids and other insects have been shown to require trace amounts of iron, zinc, manganese and copper, which are essential constituents of many of the enzymes that catalyse metabolic processes. The simplest substances which will satisfy these needs may vary to some extent from one insect species to another; but in general the requirements are not very different from those of vertebrates. *Drosophila* has been reared with ammonium salts, and cockroaches with glycine as the sole sources of nitrogen; but these insects have in the intestine a rich flora of yeasts and bacteria which were doubtless responsible for essential syntheses.

When reared under axenic conditions in the absence of microorganisms, insects are found almost always to require the same series of ten 'indispensable' amino acids as do vertebrates: leucine, isoleucine, histidine, arginine, lysine, tryptophane, threonine, phenylalanine, methionine, and valine. Nucleic acids they can synthesize. The range of hexose monosaccharides which they can metabolise varies somewhat in different species and the utilization of disaccharides and polysaccharides may be limited by the digestive enzymes present (p. 55). Some insects, such as the mealmoth *Ephestia*, must be provided with a supply of the more highly unsaturated fatty acids such as linoleic and linolenic acids, but for most species this is not necessary. On the other hand, most insects are unable to synthesize sterols in quantity and must have cholesterol or sitosterol in the diet.

Vitamin Requirements

A heterogeneous series of substances which the animal organism is unable to synthesize and which it requires in varying amounts are for historical reasons, commonly termed 'vitamins'. So far as the water-soluble or B group of vitamins is concerned, the requirements of insects are strikingly similar to those of mammals. Although there are some specific variations, thiamin (B1), riboflavin (B2), nicotinic acid, pyridoxin (B6) and pantothenic acid, are indispensable. Choline and biotin are important; inositol and p-amino-benzoic acid often of no importance. Folic acid is often required. These vitamins are mostly required to help furnish the various co-enzymes operative in intermediary metabolism: thiamin in carboxylation and decarboxylation; riboflavin for the flavoprotein involved in hydrogen transfer; nicotinic acid in dehydrogenation; pantothenic acid as a constituent of co-enzyme A; choline as a donor of methyl groups and as a component of phospholipids and acetylcholine, pyridoxin as a co-enzyme in transamination.

Fat-soluble accessory substances of the vitamin A type are not required by insects in large amounts; but 'retinene' in the visual purple (rhodopsin) of the eye is a derivative of vitamin A. If the mosquito *Aëdes* is reared for a generation on a diet containing neither vitamin A nor its precursor β-carotene, the structure and function of the retina is impaired. Addition of β-carotene eliminates these defects.

Vitamin C (ascorbic acid) which is present in quantity in the tissues of insects, they can readily synthesize themselves. But the cornleaf factor required by larvae of *Pyrausta* and *Trichoplusia* appears to be ascorbic acid. Vitamin D (calciferol) is not necessary, although as we have already seen they must have supplies of cholesterol or some related sterol. Vitamin E is essential for the production of viable offspring in the Sarcophagid fly *Agria*.

Micro-organisms and Symbionts as Sources of Vitamins

Reliable information on the vitamin requirements of insects can be obtained only if they are reared in the absence of micro-organisms. Larvae of the blowfly *Lucilia* can be grown from sterile eggs on the sterile brain of mammals; they will not grow on certain types of sterile muscle. But if the muscle is infected with suitable micro-organisms, or if a sterile extract of yeast is added to it, normal growth is obtained. The bacteria and the yeast provide vitamins (belonging to the B group of vitamins) that are present in brain but lacking in muscle. It is interesting to note that sterile blood, also, is an insufficient diet for these larvae; it can be rendered adequate by the addition of vitamin B[1].

Many insects (the bed-bug (*Cimex*), the sucking lice, the tsetse-fly (*Glossina*), the Pupipara) take no food throughout their life except sterile blood. But all these insects contain within their bodies special groups of cells (mycetomes) stuffed with supposedly symbiotic micro-organisms, which are absent from such insects as mosquitoes, fleas, &c., that take blood during part only of their life cycle. This naturally suggests that these symbionts may constitute an endogenous source of vitamin, and thus enable these insects to grow on sterile blood; indeed, it has been shown experimentally that if the louse *Pediculus* is deprived of its mycetome and symbionts, growth and reproduction are greatly impaired, and that this impairment can be made good by a single dose of pure B vitamins. The blood-sucking Reduviidae *Rhodnius*, *Triatoma*, &c., have no intracellular symbionts; but they regularly have the gut contents heavily infected with an Actinomyces which provides a source of B vitamins. When reared on blood in the absence of this micro-organism they cannot grow or reproduce.

It is clear, therefore, that when insects feed upon a sterile or restricted diet, the supply of vitamins may become a problem of some urgency. To provide an endogenous source of vitamins may well be one of the functions of symbiotic micro-organisms in other groups besides those feeding on blood. This has been proved experimentally to be the case in such beetles as *Lasioderma* and *Sitodrepa*, which do not normally need vitamins of the B group in their food. If these beetles are deprived of their symbionts (by sterilizing the surface of the egg), their vitamin requirements become the same as those of *Tribolium* or *Ptinus*, which always lack symbionts; and likewise their needs for essential amino acids.

The association of sterile food (as in the majority of plant-sucking Hemiptera) with the presence of symbionts has again been traced to this need and for other syntheses, notably of essential amino acids. If Aphids are fed on a diet containing adequate levels of vitamins, the adults pass to their developing embryos sufficient reserves of vitamins for them to complete their larval growth without further supplementation. Under these circumstances essential needs are easily overlooked.

The intracellular symbionts in the fat body of cockroaches, *Periplaneta*, and *Blattella*, can be eliminated by feeding the parent insects with aureomycin for a long period, or by exposure to a high temperature. Such insects are unable to grow on diets that are adequate for normal insects. These micro-organisms seem to be particularly important in amino acid, especially methionine, and ascorbic acid synthesis. It is not always easy to see why related forms (Cerambycid larvae, for example), with and without symbionts, should grow with equal rapidity on the same diet. But the insect and its symbionts show such wonderful mutual adaptations, that it is equally difficult to dismiss the whole phenomenon as one of harmless parasitism.

Water Metabolism

Water uptake. Another fundamental requirement in insect nutrition, and a very important one, is water. Even such an insect as the weevil *Calandra*, which is accustomed to feed in dry materials and is highly adapted to conserve the water in its body, cannot survive unless its food contains some 10 per cent. of water. In other words, the water produced from the oxidation of foodstuffs, although of great importance to the insect, is not, alone, sufficient for its needs. If the mealworm is kept in air with a relative humidity of 90 or 95 per cent., it is able, by some mechanism that is not fully explained, to absorb water from the atmosphere through the cuticle; in the larva of the flea *Xenopsylla* the critical equilibrium humidity for the uptake of water is 65 per cent. relative humidity; in the pheasant louse *Goniodes*, 50 per cent and in *Ctenolepisma* 47·5 per cent! But this property is very unusual; most insects cannot absorb water vapour even when the air is saturated.

Water loss. On the other hand, the *loss* of water (when this occurs by evaporation as opposed to excretion (p. 54)) is markedly affected by the water vapour in the air. The rate of loss, like that from inanimate bodies, has been found, within certain limits, to be more or less proportional to the saturation deficiency of the air. It is not, of course, to be expected that such a law should hold very exactly, especially at varying temperatures, because evaporation occurs chiefly from the tracheal system, the ventilation of which varies with the intensity of metabolism in the insect and so with the external temperature (p. 19).

Water and temperature control. Apart from the importance of water as a constituent of the tissues, it plays a significant part in the reaction of the insect to extremes of temperature in its environment. When exposed to high temperatures, an insect may be killed (depending upon the conditions of experiment) by the temperature itself or by the desiccation consequent upon evaporation. Provided the insect in question is sufficiently large, it will be able to lower its temperature below the fatal level (for short periods at any rate) by evaporating water from its body, and thus to exercise a limited degree of temperature control. Most insects die when their water content is reduced to about 50 per cent. of the total body weight.

Cryptobiosis. Certain insects, however, can survive in a state of suspended animation or 'cryptobiosis', after losing virtually all their water. This happens in the eggs of the Collembolan *Sminthurus*. The best known example among insect larvae is the Chironomid *Polypedilum* which will survive in a desiccated state, with its water content reduced to below 8 per cent., in the dried mud of its breeding places, and can be kept in this condition for several years. They will then withstand temperatures ranging from —190° C. (liquid air) up to 102–104° C. When moistened they quickly take up water and recover.

Cold Resistance

It is usual for insects to reduce their water content before going into hibernation and this reduction may be one factor in the increased resistance of insects to cold during the winter. But the chief factor in cold resistance is a lowering of the freezing point and undercooling point (the lower limit of supercooling) of the whole insect. One cause of this is an accumulation of glycerol in the haemolymph during diapause, where it may reach a level as high as 9 per cent. (occasionally even as high as 40–50 per cent.). At the same time there is a change of unknown nature in the haemolymph which leads to the elimination of the undefined 'nucleating agents' needed for the initiation of freezing. In some insects, notably in certain caterpillars, the body fluids may freeze solid in the winter; 'cold hardiness' then depends on the prevention of ice formation extending to the intracellular water of the tissues. Here again an accumulation of glycerol seems to be involved. Other metabolites which may contribute to cold hardiness in some insects are mannitol, trehalose and alanine.

Temperature of Insects

We have seen that at high temperatures, insects above a certain size can keep themselves cool for short periods by evaporation of water from the body, or by making use of evaporating water in the environment. At low temperatures, their bodies are often kept warmer than the surrounding air by the chemical changes going on within them. And both these methods of temperature control are used by the social Hymenoptera in maintaining the temperature of their nests or hives.

Temperature and flight. It seems that for the intense muscular activity of flight, a high body temperature is necessary: the large Sphingid moths are apparently unable to take flight immediately from rest; they first stand with the wings vibrating – shivering, as it were – until the temperature in the thorax has risen above 30° C. Only then can they fly; and during flight the temperature will exceed 40° C. If the thoracic temperature becomes unduly high the rate of heart-beat increases, the haemolymph is pumped through the thorax and heat is dissipated in the abdomen. Similar changes occur in other insects during activity; but whether insects at rest will increase their metabolism in order to maintain the body temperature is at present uncertain.

The scales and hairs which invest the body of moths and bees play an important part in insulating the insect and thus helping to maintain the high temperature during flight. In a large Sphingid moth they may serve to raise the temperature in the thorax by 8–9° C. The subcuticular air sacs play a similar role in Odonata, &c.

Radiant heat. Besides generating heat by muscular activity, many insects make use of the sun's rays to warm their bodies to the optimum temperature, orientating themselves differently according to the temperature attained: butterflies spread their wings in the sun when their temperature is too low. It has often been suggested that the tendency of insects in cold climates to be dark in colour is related to the absorption of radiant heat from the sun – though it is sometimes forgotten that this same dark colour will also favour the loss of heat during the night.

Temperature adaptation. Apart from these physiological regulations there are wide differences in the temperatures to which different insects are adapted. To quote two extreme cases: the firebrat *Thermobia* is active over the range 12–50° C.; it suffers heat injury a little above 51° C. *Grylloblatta*, which is found in the high mountains of North America, is normally active from −2·5–11·5° C.; it is irreversibly damaged by heat at 20·5° C. The physiological basis of these differences is not known.

Besides such specific differences in resistance to abnormal temperatures there are individual differences resulting from adaptation: the cockroach *Blatta*, for example, if it has been kept at 36° C., goes into a state of 'chill coma' when the temperature is lowered to 9·5°

C.; but if it has been kept, for no more than twenty-four hours, at 15°C., it does not go into chill coma until the temperature has fallen to 2° C. And there is a similar adaptation at the upper end of the temperature range.

Respiratory Metabolism

Temperature and activity. The energy metabolism of insects varies, naturally, with the degree of muscular exertion. It therefore varies enormously in different species, and is much greater in the larva than in the pupa, and greater again in the adult. It depends also upon the temperature; but a rise in temperature not only increases the resting or 'basal' metabolism, but usually leads to greater activity. Only in very exceptional circumstances, therefore, is any simple mathematical relationship likely to be found between these two factors. Curves of various types, relating temperature with oxygen uptake, notably during pupal development, have been described mathematically by some authors; while others regard such curves as being really composed of a number of straight lines which intersect at 'critical points' where new metabolic processes are supposed to supervene.

The energy metabolism of such an insect as the honey-bee, when at rest at 18° C., requires an oxygen consumption of about 30 mm^3 per gm per minute; during flight this may rise to 1,450 mm^3 per gm per minute, a 48-fold increase. Whereas in man, doing the maximum amount of work possible, the respiratory metabolism is increased only some 10 or 12 times.

Energy sources. The honey-bee, the higher flies and some other insects burn only carbohydrates during flight: they have a respiratory quotient (the ratio of CO_2 evolved to O_2 consumed) equal to one. But other insects such as Aphids and locusts, which consume mainly glycogen in the early stages of flight, rely wholly on fats for energy production during prolonged flights. And surprisingly enough, the Lepidoptera, although they feed solely on nectar, are unable to use carbohydrates directly in their flight muscles: they must first convert the sugar into fat; during flight the R.Q. is always about 0·75. In most insects, during starvation, glycogen is first consumed, to be followed by protein and fat.

The biochemical processes of energy production from stored reserves are in general outline the same as in other animals. The citric acid cycle occurs in insects as a main pathway for the liberation of the hydrogen which serves as fuel for the cytochrome system. The series of haemochromogen compounds that make up the cytochrome system were originally discovered by David Keilin when he was working with insects.

Anaerobic Metabolism

Resistance to oxygen lack. Insects are remarkably resistant to lack of oxygen. Down to a low level of oxygen tension, the precise level varying in different insects, the oxygen uptake remains unchanged. Below this level it falls off; but many species can survive for long periods in the complete absence of air. Adult *Musca* can recover from 12–15 hours without oxygen. The extra metabolism of activity very quickly ceases, and the insect becomes quiescent; the basal metabolism continues anaerobically. The type of chemical change does not alter: the unoxidized metabolites simply accumulate. Thus, by the time the insect is again restored to the air, it is greatly in arrears for oxygen; and in order to pay off this 'debt', and oxidize the lactic acid and other substances that have accumulated, its rate of oxygen consumption remains above the normal level for a long time (Fig. 14, B). During the anaerobic period, the blood naturally becomes more acid, the pH falling, in the case of the grasshopper, from 6·8 to 5·8, and the carbon dioxide capacity is reduced; consequently, at very low tensions of oxygen, there may be a liberation of combined carbon dioxide. This type of temporary anaerobic metabolism is a normal occurrence in the life of *Gasterophilus* (parasitic in the stomach of the horse) which can survive as long as 17 days in the complete absence of oxygen; experimentally, it occurs in insects of all kinds.

The anaerobic production of energy by this process of 'glycolysis' can be utilized only in certain of the less active muscles of insects. It will not serve to maintain flight. The flight muscles are very richly supplied with tracheoles; their metabolism is strictly aerobic. They build up almost no oxygen debt, and can function only so long as the lactate and particularly the α-glycerophosphate produced by glycolysis can be completely and directly oxidized.

Metabolism and Chemical Changes during Growth and Metamorphosis

Curve of oxygen consumption. The curve of oxygen consumption during the pupal stage of insects usually follows a more or less U-shaped course; it falls first and then rises again before emergence (Fig. 14, A). The level of oxygen consumption at any moment

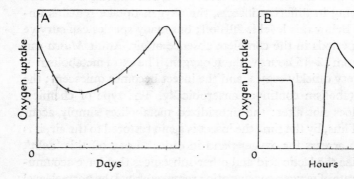

Figure 14. Ordinates: rate of oxygen uptake; abscissae: time

A, oxygen uptake during the whole pupal stage of $9\frac{1}{2}$ days in *Galleria*, Lep. (after Taylor and Steinbach); B, oxygen uptake in grasshopper, *Melanoplus*. During the 75 minutes that the curve coincides with the base line, the insect was submerged in water (after Bodine)

represents the sum of the oxygen required by all the different tissues. What may be the chief organs or the chief activities consuming oxygen at the different points along this curve have not been fully worked out. But since the quantity of respiratory enzymes present will run more or less parallel with the respiratory requirements it is not surprising that the cytochrome system and the substrate dehydrogenase systems follow a similar curve. An increasing rate of metabolism may be evident in the developing pupa at a time when almost no mechanical work is being done. The oxygen concerned is required for endothermic synthesis, and perhaps mainly for protein synthesis, which is an integral part of the growth process and which demands a large amount of energy.

Chemical changes. The changes in the total composition of the insect body during metamorphosis naturally vary in different groups, with the different types of morphological change. As a rule, the large stores of *glycogen* accumulated towards the end of larval life (these may exceed 33 per cent of the dry weight of the body in the six-day-old bee larva) fall steadily, almost reaching zero in the developed adult. The *fat* decreases rapidly at first, then more slowly (during the period of depressed oxygen consumption) and then rapidly again before emergence. Fat does not usually fall so low as glycogen, but only a small percentage of the original store persists in the adult. In the pupa of the Muscid *Lucilia*, it appears to be the unsaturated fatty acids which are consumed, the saturated acids remaining more or less unchanged. The *protein* is sometimes largely used in the formation of the cocoon (the silkworm uses half its body protein for this purpose); the remainder is extensively broken down during the early part of pupal life, the non-protein nitrogen, particularly the amino acids, increasing. With the growth of the new tissues, the process is reversed, and the amino acids fall (p. 38). The inorganic phosphate follows the same course as the amino-acids.

These changes, and the corresponding changes in other constituents and in the respiratory quotient, represent, of course, only the summation of the processes of synthesis and breakdown that must be going on throughout metamorphosis. At times it must happen that these two processes will exactly balance in respect to a particular substance over a given period, and then chemical analysis will reveal no change at all. It is clear, therefore, that such analyses of the whole animal are of rather limited interest; or rather, are impossible to interpret intelligently, until they have been correlated with the histological sequence of events. Conversely, intricate cytoplasmic changes are described in the fat body and blood cells during meta-morphosis, the chemical significance of which is incompletely known. Ribonucleic acid, proteins, fats, and carbohydrates are being synthesized, particularly by the fat body cells. In what form they are liberated for supplying the growing imaginal tissues is uncertain.

Some Chemical Products of Insects

Silk, wax and shellac. Insects produce a wide range of organic compounds for many of which they have found biological uses, and some of which have become articles of commerce. Prominent among these are the fibrous protein 'silk', from the commercial silkworm; the 'waxes' of the honey-bee and other insects, which are closely related to the water-proofing waxes of the cuticle (p. 5); the mixture of resinous substance, hard wax, and pigment which forms the 'lac' of commerce, the product of dermal glands in various Coccidae. The resinous product 'shellac' is a complex polyester of varied compounds carrying carboxyl and hydroxyl groups.

Venoms and defensive secretions. Hymenoptera Aculeata produce 'venoms' which range from the formic acid of Camponotine ants, to the toxic protein of bee-venom. Lepidoptera, Coleoptera, and Heteroptera produce 'defensive secretions' of the utmost variety. Many contain quinones, or unsaturated aldehydes; salicylaldehyde, hydrocyanic acid, and histamine may be secreted. Caterpillars, notably of Papilonidae, will take up and discharge the poisonous compounds such as alkaloids or cardiac glycosides in certain plants (Aristlochiaceae, Umbellifera, &c.).

Pheromones. Other substances, many of them not yet identified chemically, serve as 'pheromones' and act as chemical messengers in relation with other individuals of the same species. They may serve as 'alerting substances', 'sexual attractants', 'aphrodisiac scents', and other forms of 'social communication'. One of the best known is the 'queen substance' secreted by the mandibular glands of the queen bee, which has been identified as 9-oxodecanoic acid and is used by the workers to recognise the continued presence of the queen, and by the queen during the mating flight as a sexual attractant for the drones.

Pigment Metabolism

Nature and origin of pigments. We have already seen that in the course of metabolism pigmented substances may arise as waste products (p. 65). Pigments may also be absorbed with the food and accumulate within the body: the flavone pigment in the wing-scales of the butterfly *Melanargia*, for example; and the carotinoids in the blood of the potato beetle *Leptinotarsa*, and in the epidermal cells of the bug *Perillus* which preys upon it. But it is remarkable that apparently similar pigments in more or less related insects may be quite unrelated chemically: thus, the common red and orange pigments of many Hemiptera may be carotinoids from the food, anthocyanin-like or flavone-like pigments, perhaps also from the food, pterins synthesized along much the same lines as uric acid, or ommochromes derived from the metabolism of the amino acid tryptophane; and the Pieridae may use their pterin colours to mimic other butterflies the chemistry of whose pigments is quite different.

Green colours. Similarly with the green pigments: in many cases these are synthesized by the insect itself, and have no relation to the chlorophyll of plants, even in phytophagous species; for their development is unaffected (in *Carausius*, in locusts, and in many caterpillars) by rearing them upon food containing no pigment; nor have they the chemical properties of chlorophyll (p. 66); they are commonly mixtures of the blue biliverdin with some yellow chromoprotein containing xanthophyll or carotene. The sexual differences in the blood of many caterpillars, which have been ascribed to differences in the utilization of plant pigments, are probably due rather to differences in the composition of the haemolymph proteins.

Melanin. The other type of insect pigment is the insoluble black substance melanin, which is formed in the presence of oxygen by the action of an oxidizing enzyme upon some colourless aromatic precursor, the 'chromogen'. Melanin-producing enzymes are generally diffused throughout the blood and tissues and secretions of insects, particularly near the time of pupation when much black pigment is often formed; the chromogens are usually deposited in restricted areas, and so determine the colour patterns that develop.

The melanogenic enzymes can act upon a great variety of chromogens: tyrosine (monoxyphenylalanine), 'dopa' (dioxyphenylalanine) and various other polyphenols; the complex of enzymes concerned being termed tyrosinase. In the tissues and in the cocoons of certain Lepidoptera (*Samia cecropia*, *Eriogaster lanestris*, &c.) and in the elytra of *Melolontha*, the spontaneously oxidizable 'dopa' has been demonstrated; and in the mealworm *Tenebrio*, dioxyphenylacetic acid has been proved to be one of the chromogens present. It is, of course, possible that more than one chromogen may be concerned in a given insect; and in any case it is not easy to separate this process of melanin formation in the cuticle from the closely allied process of quinone tanning of protein that is responsible for the hardening of the cuticle which is going forward at the same time. Experimentally, the quantity of melanin formed can be widely varied by changes in the external conditions, notably the temperature. In some cases (*Habrobracon*, Hym.), these effects may be transmitted to the offspring up to the second generation.

Melanin is incorporated in the substance of the cuticle; other brown pigments (insectorubins or ommochromes) are deposited as granules in the epidermal cells. Some pigments are in solution in the blood or in the fat droplets of the cells (lipochromes), many appear as inclusions in the epidermis or other tissues or between the layers of the wing scales.

Colour change: physiological. In some insects the coloration may be strikingly influenced by external factors. Sometimes, notably in the stick insect *Carausius*, such changes are reversible – the so-called 'physiological colour change'. There are no discrete chromotaphores in these cases, there is simply a migration of pigment granules, a clumping or spreading, in all the epidermal cells. In *Carausius* this change may be brought about by a great variety of stimuli: temperature, narcotics, osmotic changes in the blood, mechanical pressure, all of which act directly upon the epidermal cells; and optical stimuli and humidity, which act through the eyes and tracheal system respectively, and then, by way of a centre in the brain, lead apparently to the secretion of a hormone into the blood, which determines the pigment movements. These pigment changes follow, also, a diurnal rhythm (the insect is dark by night and pale by day) which, though determined primarily by illumination, may persist for several weeks in complete darkness.

Colour change: morphological. In these same insects, if the factors or stimuli already enumerated continue to act over long periods, the amounts of the different pigments formed may be permanently changed; and this may be effected, also, by nutrition, particularly by the moisture of the food. This permanent effect is termed 'morphological colour change'; but when brought about by optical stimuli, at any rate, it appears again to be produced, like the temporary change, by way of the nerve centre and the internal secretion. The co-ordination of these mechanisms may lead to remarkable resemblances to their background in various Orthoptera.

As was shown by Poulton many years ago, larvae and pupae of certain Lepidoptera also possess wonderful powers of acquiring the tone or coloration of their background. This phenomenon has been analysed most carefully in the case of *Pieris* pupae. In these the quantity of pigment deposited in the cuticle appears to be determined by the specific quality of the light received into the eyes of the pupating larva, acting through nervous and hormonal centres situated in the head and thorax.

Another very striking colour change is that in locusts, the solitary phases of which are quite differently pigmented from the gregarious phases. The gregarious type of coloration can be evoked by crowding the young forms together; and here again the immediate cause of the colour change is probably the hormone secretions, perhaps in the corpus allatum or corpus cardiacum, controlled by the brain.

Light Production

A certain number of insects are luminous; they liberate energy in the form of light during their metabolism. This property probably arose as an accidental accompaniment of the metabolism that occurs in the fat body. In the most primitive luminous insects (Collembola) the fat body throughout the insect is luminous; and in the most highly evolved fire-flies (Coleoptera) the light-producing organ consists of specialized fat-body cells lying beneath a transparent window in the cuticle with other specialized cells loaded with uric acid granules as a reflector beneath. But in the luminous Mycetophilid larva *Bolitophila* it is the Malpighian tubes which form the luminous organ.

Light is produced when an enzyme 'luciferase' acts upon a substrate 'luciferin'. The quantity of heat set free in this reaction is very small; at least 98 per cent. of the radiant energy appears in the form of light, usually limited to the greenish-yellow region of the spectrum. In the fire-flies and glow-worms the production of light, whether as a steady glow or as a succession of flashes, is under the control of the central nervous system. The mechanism of control is uncertain: it is sometimes thought to be effected by a system of minute sphincters in the tracheal endings which control the access of oxygen to the light-producing cells; but it seems more probable that a neurohumour (resembling adrenalin) is liberated at the nerve endings and acts upon the light-producing cells. In the luminous beetles the light certainly serves as a mating signal.

BIBLIOGRAPHY

BEARD, R. L. *Ann. Rev. Entom.*, **8**, (1963). 1–18 (insect toxins and venoms: review)

BROOKS, M. A. *Symp. Soc. Gen. Microbiol.*, **13**, (1963), 200–231 (symbiosis in insects: review)

BURSELL, E. *Physiology of Insecta* I (Morris Rockstein, Ed.) Academic Press, New York, 1964, 323–361 (temperature and humidity relations)

CHEFURKA, W. *Physiology of Insecta* II (Morris Rockstein, Ed.) Academic Press, New York, 1964, 582–768 (intermediary metabolism)

CROMARTIE, R. I. T. *Ann. Rev. Entom.*, **4**, (1959), 59–76 (insect pigments: review)

EDNEY, E. B. *The Water Relations of Terrestrial Arthropods*, Cambridge University Press, 1957, 109 pp

EISNER, T. and MEINWALD, Y. C. *Science*, **153**, (1966), 1341–1350 (defensive secretions of insects: review)

GILBERT, L. I. *Adv. Insect Physiol.*, **4**, (1967), 69–211 (lipid metabolism in insects: review)

GILMOUR, D. *The Biochemistry of Insects*, Academic Press, New York, 1961, 343 pp.

HABERMANN, E. *Science*, **177**, (1972), 314–322 (bee and wasp venoms)

HOUSE, H. L. *The Physiology of Insecta* II (Morris Rockstein, Ed.) Academic Press, New York, 1964, 769–813 (insect nutrition)

MCELROY, W. D. *The Physiology of Insecta* I (Morris Rockstein, Ed.) Academic Press, New York, 1964, 463–508 (light production)

SACKTOR, B. *The Physiology of Insecta* II (Morris Rockstein, Ed.) Academic Press, New York, 1964, 484–580 (respiratory metabolism in muscle)

SALT, R. W. *Ann. Rev. Entom.*, **6**, (1961), 55–74 (insect cold-hardiness: review)

WIGGLESWORTH, V. B. *Tijdschrift Ent.*, **95**, (1952), 63–68 (symbionts in blood-sucking insects: review)

——*The Principles of Insect Physiology*, 7th Edn. Chapman and Hall, London, 1972, 513–526 (nutrition); 513–622 (chemical transformations and pigment metabolism); 622–636 (respiratory metabolism); 663–690 (water and temperature relations); 451–455 (light production)

7 Growth

The phenomena of growth and development in insects are, in broad outline, like those of higher animals. At the posterior pole of the egg there may an 'activation centre' where, it is suggested, some active substance is liberated which sets in motion the subsequent processes of development. In all insect eggs there can be recognized in addition a 'differentiation centre', lying in the anterior third of the germ-band region, which acts as the leading point in all the succeeding stages of embryonic development.

These 'centres', the nature of which is quite unknown, are often active in the cortical plasma of the egg before its division into cells, and they cause this to become transformed into an invisible 'mosaic' of regions 'determined' or committed to form a given part of the embryo. The cleavage nuclei which migrate through the yolk and settle down at the surface to form the cellular blastoderm are thus induced to contribute to the formation of different organs according to the region in the plasma where they chance to come to rest. Those which arrive outside the germ-band zone form only the extra-embryonic blastoderm.

Once differentiation has begun, development, co-ordinated by some unknown mechanism, proceeds until the insect hatches. Then follows a period of growth ending in maturity and reproduction.

Growth Ratios

Allometric growth. As in other animals, there is often a change in the proportions of the body as it increases in size; and when this is the case, the disproportion generally follows the allometric law, that is, the logarithm of the dimension of the part is proportional to the logarithm of the dimension of the whole: $y = Kx^a$, where x is the dimension of the whole, y the dimension of the part, a the 'growth coefficient' and K another constant – each part increasing in size geometrically at a rate peculiar to itself.

Discontinuous growth. Upon this general background are superimposed some special phenomena, conditioned by the make-up of the insect; and it is in these that we are mainly interested. In the first place, the cuticle is so constructed that once formed it cannot grow in surface area. If it is soft and much folded it may *stretch* gradually, and then the insect may increase continuously in length as it does in weight. This happens in the larvae of flies (e.g. *Drosophila*) and Lepidoptera (e.g. *Galleria*); but even in these insects there are hard regions of the cuticle which can grow only by replacement at the time of moulting. The growth of such parts is, therefore, discontinuous; and when the greater part of the insect is encased in a rigid cuticle, the growth in length of the whole body is likewise discontinuous, and takes place in a series of steps.

Growth ratios. It was observed long ago by Dyar, working with the larvae of Lepidoptera, that the linear dimensions of the hard parts increased at each moult by a constant ratio, usually of the order of 1·4 (Dyar's law). More recently, it has been shown that various hemimetabolous insects (*Sphodromantis*, *Carausius*) approximately double in weight from one moult to the next, while their linear dimensions increase 1·26 (i.e. $\sqrt[3]{2}$) times. This growth factor of 2 was first demonstrated by Przibram and Megusar; it is perhaps best regarded simply as a special case of Dyar's law; but it is sometimes referred to as 'Przibram's factor'. The physiological basis proposed by Przibram to explain his factor was that at each moult (or at each latent division) every cell in the insect body divides into two. But what histological evidence exists is definitely opposed to this improbable hypothesis. For in the Muscidae, almost the whole of larval growth is accomplished by increase in size of the cells without multiplication, and in hemimetabolous insects there is much cellular breakdown and reconstruction besides the irregularly distributed cell divisions.

Growth curves. We have seen that at the time of moulting, the insect enlarges its surface area by swallowing air or water (p. 11). In the latter case (*Aeschna, Notonecta*, &c.) the resulting curve for growth in weight is peculiar and shows an abrupt rise immediately after moulting, followed by a more or less level period during which the ingested water is gradually incorporated as true growth (Fig. 15, B). Equally curious is the growth curve of the blood-sucking bugs (*Cimex, Rhodnius*) which take only a single gigantic meal during each moulting stage, so that the weight curve shows a series of acute peaks gradually rising to a higher level (Fig. 15, C).

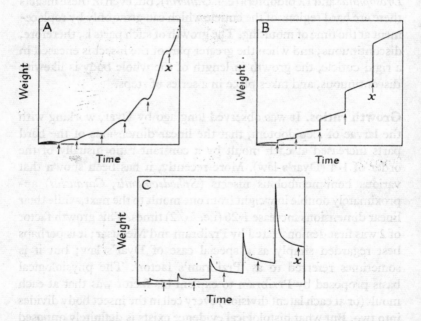

Figure 15. Growth curves; semischematic; A, *Dixippus* (modified after Teissier); B, *Gerris*, Hem. (modified after Teissier); C, *Rhodnius*, Hem. (Wigglesworth)

The arrows indicate the time of moulting; the insect becomes adult at x

Moulting

Definitions. The terms moulting and ecdysis in insects referred originally to the casting of the old skin. But both terms have come to be used in two senses: (i) for the entire process of growth which leads up to the next stage, or stadium; (ii) for the final act of shedding the cuticle. Some authors limit the use of the term moulting to the former sense and ecdysis to the latter, but there is no uniformity in this. Two other useful terms have been introduced by Hinton: 'apolysis' for the separation of the epidermis from the old cuticle, the precise timing of which is often difficult to determine; and the 'pharate state' for the period between apolysis and ecdysis when the new cuticle is in process of formation below the old.

The chief interest of insect growth centres round the phenomenon of moulting or ecdysis. This is essentially a process of growth; for it is the only means by which the hard parts can increase in size (p.8). There is no doubt, therefore, that growth is one factor that determines moulting. But in some insects (*Tineola, Tenebrio*) moulting can occur repeatedly during starvation – and without any enlargement: *Tineola* may sometimes develop a sort of 'moulting fever' and cast the skin as many as eight times in two or three days; and adult Thysanura will continue to moult without change in size or organization.

Brain hormone. There must therefore be some other factor at work. This factor has proved to be a hormone circulating in the blood For example, the blood-sucking bug *Rhodnius* takes only a single feed in each of its five larval stages; moulting takes place at a definite interval (varying from 10 to 20 days in the different instars) after each meal. If the recently fed insects are decapitated before a certain 'critical period' after feeding they fail to moult, although some of them may survive for more than a year. But if an insect decapitated soon after the critical period is connected by means of a capillary tube to an insect decapitated one day after feeding, so that the blood passes from the one to the other, the second insect is caused to moult.

A similar effect, of decapitation preventing pupation, was observed by Kopeč (1917) in the caterpillar of the Gipsy moth *Lymantria*; and since he found that pupation could be prevented by removal of the brain, though not by section of the nerve cord, Kopeč suggested that the brain was secreting the hormone in question. In *Rhodnius*, and in all other insects so far examined, there are groups of 'neurosecretory cells' in the dorsum of the brain, and the implantation of these cells into the decapitated insect will induce moulting.

Moulting hormone. But this factor from the brain is not the 'growth and moulting hormone' which acts upon the tissues, for implantation of the neurosecretory cells will not induce moulting in the isolated abdomen. The brain factor seems merely to activate another gland of internal secretion, the 'ventral gland' in the head of some insects, the 'thoracic' or 'prothoracic gland' in others, and it is this structure (which had been observed by Lyonet in the larva of *Cossus* as long ago as 1762) that is the source of the hormone that induces moulting. That has been proved experimentally in Lepidoptera, in Diptera (where the gland in question forms part of the 'ring gland' of Weismann), in *Rhodnius*, in the cockroach and in other insects. When the adult stage is reached the thoracic gland degenerates and the insect does not moult again – except in the primitive Thysanura in which moulting continues in the adult and the ventral gland persists.

As to the stimulus which causes secretion of this hormone: there is evidence that the stimulus to moulting in *Rhodnius* is the stretching of the abdomen by the meal; and section of the nerve cord in the prothorax inhibits moulting like decapitation. This suggests that moulting is started in *Rhodnius* by a nervous stimulus to the brain. But this cannot, of course, be a general explanation, applicable to those insects which moult repeatedly though starved. In insects such as locusts, which feed continuously, it seems to be the act of swallowing that provides stimuli which go via the frontal ganglion to the neurosecretory cells of the brain and thus maintain a continuous production of the moulting hormone by the ventral or prothoracic glands.

In *Rhodnius* there are histological signs of secretion by the neurosecretory cells beginning immediately after feeding and lasting 6–8 days; a second phase of release accompanies ecdysis. In silkmoths an 'eclosion' hormone liberated from the brain has been shown to set in motion the specific behaviour pattern which leads to ecdysis.

Metamorphosis

Thus far we have considered only the factors which control growth and moulting. But it is characteristic of insects that moulting may be associated with a more or less striking change in form or 'metamorphosis'. In some groups of insects the larva does not transform directly into the winged adult but into an intermediate, non-feeding stage, the pupa, which then moults again to produce the perfect insect. This pupal stage, however, is not a necessary feature of insect metamorphosis. The change in form from the larva to the adult may be just as great in such insects as dragon-flies (Odonata), whiteflies (Aleurodidae), or may-flies (Ephemeroptera) which have no pupal stage, as it is in the alder fly (Megaloptera) with a well-defined pupa.

Imaginal discs. In the larva of some insects, notably in Diptera and Lepidoptera, many of the organs of the adult are already recognizable as clusters of embryonic cells, the so-called 'imaginal discs' of Weismann. But these likewise are not a necessary feature of metamorphosis. In many insects which show a striking metamorphosis, such as dragon-flies, they do not occur; and they are absent, for example, from the abdomen of caterpillars which is transformed into the totally different structure of the butterfly.

The essential feature of insect growth is the existence within the epidermis of a latent capacity to develop into several organisms of widely different form. The potentiality to form the pupa or the adult remains latent within the functioning cells of the larva; metamorphosis consists of the realization of these latent genetic capacities for the growth and differentiation.

Juvenile hormone. The control of this process is well seen in *Rhodnius*. During the first four larval stages the corpus allatum (a small endocrine gland lying just behind the brain) secretes the so-called 'juvenile hormone' or 'neotenin' which ensures that the larval organism is again produced when the insect moults. In the fifth larval stage the corpus allatum no longer secretes the juvenile hormone so that when moulting is brought about by the 'growth and moulting hormone' the latent imaginal organism is able to develop and metamorphosis takes place. If the corpus allatum is removed from the young stages of insects, precocious metamorphosis occurs: diminutive silkmoths weighing only a few milligrams have been produced in this way. Or, conversely, if the corpus allatum from a young insect is implanted into a late stage, giant larvae can be produced. The juvenile hormone reacts directly with the epidermal cells, for if only a little of the hormone is produced from an implanted gland, it may result in a tiny patch of larval cuticle over the implant developing in an otherwise normal adult.

There is evidence that in such insects as Lepidoptera, with a pupal stage, the larva develops when much juvenile hormone is present, the pupa when only a little is present, the adult when the hormone is absent altogether. In parts of some insects, such as the abdominal integument in *Rhodnius* or *Oncopeltus* or in Lepidoptera, if the adult is caused to moult once more by supplying it with the moulting hormone and if at the same time it is supplied with abundant juvenile hormone, it may show a partial reversal of metamorphosis with the partial development of larval characters again.

Prothetely. It would seem that the gradual changes in form that occur in the successive stages of insects, as well as the more striking changes at metamorphosis, are likewise controlled by a delicate balance in the timing and concentration of the secretions from the corpus allatum and thoracic gland. Sometimes this balance is upset within the intact insect, particularly in species hybrids or in insects reared under abnormal conditions, and then monsters intermediate between larva and pupa, or between pupa and adult, are produced. This phenomenon is termed 'prothetely' or 'metathetely'. It may sometimes be induced by microbial or virus infections in insects.

Chemistry and Action of Insect Growth Hormones

Brain hormone. The secretion of the neurosecretory cells in the dorsum of the brain is made up of minute spheres of protein (100–300 nm in diameter) which pass along the axons that end blindly in the 'corpus cardiacum' lying just behind the brain. The active material is believed to be liberated from the corpus cardiacum; it is probably a protein, or polypeptide. It certainly acts upon the thoracic gland and causes this to secrete the moulting hormone which is named 'ecdysone'.

Ecdysone. Ecdysone has been isolated and crystallized in pure form from larvae of *Calliphora* and from pupae of the silkworm: it is a steroid closely related to cholesterol (Fig. 16); a second, more active, form has one additional hydroxyl group and is named

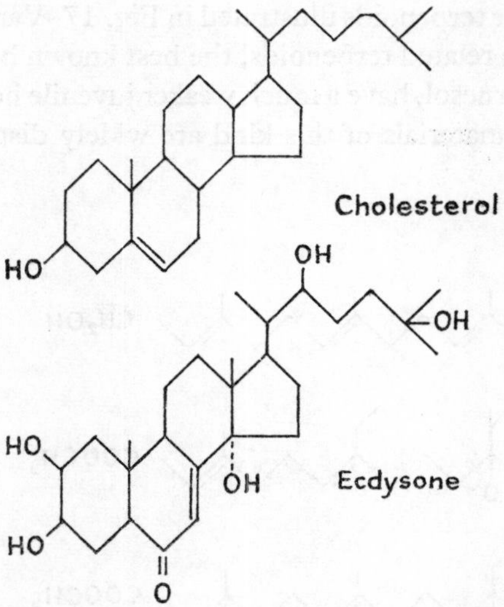

Figure 16. Comparative skeleton formulae of cholesterol and ecdysone, showing the large number of hydroxyl groups in ecdysone, which are responsible for its solubility in water

ecdysterone. Ecdysone exerts its effect directly upon those cells concerned in growth and moulting; it 'activates' them and stimulates them to renewed protein synthesis. It may be that it acts directly upon those gene loci in the chromosomes that are concerned in growth; it will induce characteristic 'puffing' in the giant polytene chromosomes of Diptera and these 'puffs' are believed to be the sites of formation of the messenger RNA required for protein synthesis. Exposure of *Rhodnius* to a high temperature of 36°C arrests growth without interfering with other body functions; this effect appears to be an arrest of protein synthesis.

Juvenile hormone. The juvenile hormone, besides its control of metamorphosis, is concerned in the reproductive activity of both sexes (p. 122). It accumulates in considerable quantities in the adult male of the giant silk-moth *Hyalophora cecropia*, and if the fats are extracted from the abdomen of this moth they appear as a yellow oil rich in juvenile hormone activity. This material served as a source for the isolation of the hormone which has now been identified as a mixture of the terpenoids illustrated in Fig. 17. Various substances, most of them related terpenoids, the best known being the familiar metabolite farnesol, have a much weaker juvenile hormone action in insects; and materials of this kind are widely distributed in small

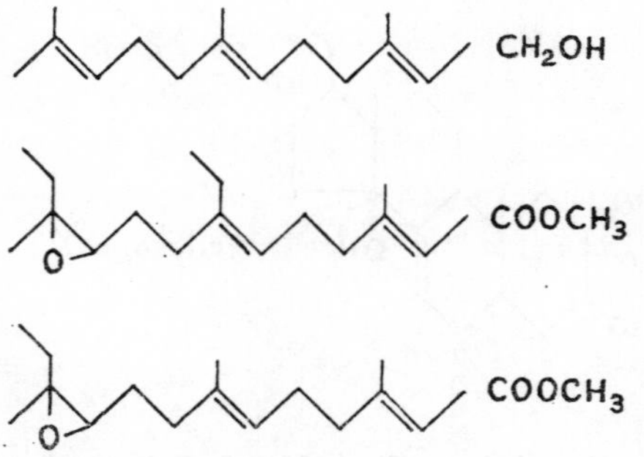

Figure 17. Skeleton formulae of the sesquiterpene farnesol (I) and the two terpenoid juvenile hormones in the cecropia silkmoth (II and III)

amounts in animals and plants. The mode of action of the juvenile hormone will be considered in relation with 'polymorphism' (p. 110).

Histolysis and Histogenesis

In the processes of growth and in the determination of form in insects the effective tissue is the epidermis. There is the utmost variation in the behaviour of the internal organs. Even in the extreme case of metamorphosis in the Diptera, many larval organs (Malpighian tubules, certain muscle groups) are re-modelled without much change to form those of the adult. Other organs undergo extensive breakdown (histolysis) and new organs are built up from the embryonic cells of the imaginal discs (histogenesis). There has been much controversy about the physiology of this process. It is now clear that when metamorphosis begins certain of the larval cells die and disintegrate (autolysis), but the physiology of this process is not known. The phagocytic haemocytes are some-times actively concerned in removing this disintegrating tissue and making it available to the growing organs of the adult, but there is no clear evidence that the haemocytes initiate this process. In *Calliphora* a change in the haemocyte population, brought about by the moulting hormone, is one of the first indications of the onset of metamorphosis.

The co-ordination of all these processes to produce the adult form is the problem of orderly growth which is at present quite unsolved. The potentialities of form seem generally to be inherent in the tissues themselves (wings, eyes, integument, &c.) and to be more or less independent of the other organs in the body. Though the develop-ment of the muscles at metamorphosis seems often to depend upon their connexion with the central nervous system.

Regeneration

The general obscurity of the factors controlling growth is well illus-trated by the phenomena of regeneration. In the adult insect the capacity to heal wounds, with the formation of new cuticle, persists;

but there are no extensive powers of regeneration. In the pupa the same is true; if the developing wings in the pupa of *Tenebrio* are cut short, they round themselves off and approximate to the normal form, but there is no true regeneration of tissue. If the wing germs of Lepidoptera are mutilated in the last larval stage, the damaged areas of pattern are not made good; whereas if this is done in the preceding instar, wings of normal proportions but reduced size regenerate; and if the germ is divided, reduplicated wings result. At a still earlier stage, the wing germs can be extirpated completely and are formed anew from the hypodermis.

Clearly throughout development, as differentiation proceeds, there is a progressive loss of totipotence and regenerative power; in the pupa everything is finally determined except for relatively slight changes in colour pattern. In hemimetabolous insects, if the appendages are removed in the early larval stages, they show progressive regeneration at succeeding moults, and may attain almost normal proportions by the time the insect becomes adult. Here again there is a progressive loss in regenerative power as the insect grows older, which is due partly, but not entirely, to the reduction in the number of moults in the older insects, and the corresponding reduction in the opportunities for growth.

In all these cases, the removal or partial removal of an organ results in more or less accelerated growth at the site of the wound. In the case of a limb, the new growth is most rapid if removal is effected at the level where autotomy (the spontaneous separation of an appendage) occurs naturally; but it is not necessarily connected with the property of autotomy, for there is regeneration if appendages are cut through above or below this level, and there is regeneration in appendages which cannot autotomize.

The relation between regeneration and general growth and moulting is puzzling. If a leg is removed from *Blattella* at a very early stage in the moulting cycle, moulting is arrested until regeneration of the limb is complete. It seems that the moulting hormone is not necessary for regenerative growth. Indeed in the larva of *Ephestia* (*Anagasta*) regeneration of a wing rudiment can take place after the brain and thoracic gland have been removed so that moulting hormone cannot be secreted.

The influence which moulds the outward form of the new appendage seems to reside in the epidermis itself: regeneration is unaffected by extirpation of the corresponding ganglion; though when this is done, neither muscles nor nerves are formed within the new appendage. In some cases, regeneration is 'heteromorphous': in *Carausius*, if the antenna is cut through at the level of the flagellum, it regenerates as an antenna; if it is amputated through one of the two basal segments it commonly regenerates as a leg, complete with tarsal claws. But though this tendency exists, there is no absolute relation between the level of section and the type of regeneration.

Thus, the phenomena of regeneration, like those of normal growth, can be defined by certain laws – but they are laws of growth and of regeneration; they cannot yet be expressed in the simpler terms of physiology, far less in terms of physics and chemistry.

Determination of Sex

Sexual dimorphism. The genetic theory of sex determination has been admirably described by F. A. E. Crew in *Sex Determination*, Methuen. This theory, which is based very largely on studies of insects, will explain the determination of sex both in ordinary sexual reproduction and in parthenogenesis. It can probably be accepted that, in insects, both primary and secondary sexual characters are determined by the genetic make-up of the body cells – the secondary characters are not evoked by internal secretions from special glands, either in the gonads themselves or elsewhere. For (i) removal or transplantation of the gonads at any stage of life is without effect upon the secondary sexual characters; (ii) where the genetic constitution, in respect to sex, of some body cells differs from that of others, a sexual mosaic or gynandromorph results, parts of which are entirely male and other parts entirely female; and (iii) if the wing germs of Lepidoptera of one sex are transplanted into the other sex, their sexual characters are not influenced by their new environment.

Intersexes. But in insects, as in other animals, the genetic sex can be overridden by other factors, and it is these other, physiological, factors in which we are interested here. The phenomena are most conveniently described in terms of Goldschmidt's well-known theory of intersexes. Leaving aside the genetical analysis, this theory supposes that sex is determined by a quantitative balance between male determining and female determining factors within the nucleus of every cell. By analogy with the hormones of the animal body, these competing influences are often pictured as being *chemical* in nature (although, of course, they might be of some quite other kind) – that is, as hormones working inside the cell.

In the normal female the quantitative influence of the female determining factor outweighs that of the male; but the male factor is still there, and under certain circumstances its effect may become manifest. Thus, the balance between these factors, though normally determined according to the laws of genetics, may (according to those who accept this interpretation of the facts) be upset in several ways. (i) By crossing races in which the sex determining factors are of unmatched strength. Hence 'intersexes' result, the morphological characters of which are best explained by supposing that, owing to a difference in the rate at which the male and female factors come to exert their influence, the insect has developed up to a certain point as one sex and then 'switched over' to complete its development as the other. (ii) By abnormal external temperatures, which are supposed again to upset the rate at which the two factors work, in such a way that the latent factor, which would normally be suppressed, is able to exert its specific effect, so that intersexual forms result. (iii) By the action of internal parasites – notably the effect of *Stylops* upon certain Hymenoptera. This effect was formerly called 'parasitic castration'; but it is more than the mere assumption of a *neuter* condition, it often involves a partial *reversal* of sex: females acquire male characteristics, and *vice versa*. In other words, the *latent* sex factor is again manifesting itself. This result seems to be due to starvation, which may lead to lack of some specific nutrient factor, for it occurs only in those Hymenoptera which receive a fixed ration to last the whole of larval life, and not in those social forms which are fed according to their individual needs; and the effects are exaggerated if more than one parasite is present.

Determination of Other Characters

Polymorphism. There are other types of dimorphism or poly-morphism which have much in common with the phenomenon of sex reversal. Certain parasitic insects exist in several forms with strikingly different morphology: the type which appears depends upon the size of the host or the species of host in which they are reared. Social insects, notably termites, exist in a variety of different forms or castes which are likewise determined by the environmental conditions and, indirectly, by the needs of their societies. The reproductive forms of termites prevent the appearance of other reproductive forms by a chemical factor, or 'pheromone', given off in their excreta; if the young workers do not receive this secretion certain of them develop into 'supplementary reproductives', or neotenic adults. The appearance of the soldier caste is probably regulated in the same way. The normal female, or 'queen', of the honey-bee is produced by special feeding of female larvae, which receive a liberal supply of 'royal jelly' from the salivary glands of the nurses throughout their growth. 'Solitary' and 'gregarious' phases of locusts are induced by the sparse or crowded conditions of life in the growing larvae (p. 93). Likewise winged partheno-genetic Aphids are produced by overcrowding (the so-called 'group effect') and the wingless parthenogenetic forms by solitary development. In all these cases the form of the body is influenced by factors, presumably of a chemical nature, acting upon latent morphogenetic systems. To that extent they are comparable with the phenomenon of metamorphosis.

Mosaics. For the most part, bodily characters are determined genetically. When multinucleate eggs are fertilized by several spermatozoa carrying different genes, mosaics result in the adult insect, exactly comparable with the sexual mosaics or gynandro-morphs. But characters, ordinarily determined by the genetic constitution, can sometimes be elicited by changes in external conditions during development. According to the theory of Gold-schmidt, these effects can be explained, as the occurrence of intersexes can be explained, by supposing that many morphological variations are induced by changes in the velocities of particular processes of development. Such changes in time relations may be brought about, on the one hand, by changes in the genes, or on the other, by alterations in the temperature or other factors, which influence one process more than another.

Gene switching. Nowadays it is more usual to picture the different morphological characters in different individuals of a polymorphic species, or at different stages in the life cycle of a given individual (as in metamorphosis) or in the different parts of the body, as being the result of activation of different components of the gene system. In some way that is not yet understood the juvenile hormone, the various inductors responsible for differentiation, and the factors responsible for the appearance of the different forms in insects subject to 'environmentally controlled polymorphism', all have the effect of bringing into action particular elements of the gene system: gene function is said to be 'switched'. Other types of polymorphism are due to genetic differences between individuals; the balance between the numbers of each type being determined by natural selection.

Pattern and Gradients

There is evidence for the existence of 'gradients' of an unknown nature which influence the structural pattern of the cuticle of insects. These are sometimes considered to be gradients of concentration of some chemical substance. The direction of the gradient controls the orientation of hairs etc.; and a particular 'level' in the gradient may control the type of cuticle laid down, that is, it may determine a localized gene function.

Arrested Growth or Diapause

Quiescence. Growth and development in insects may be temporarily arrested at any stage in the life cycle by adverse conditions of many kinds. Low temperature, drought, starvation, lack of vitamins, or other essential substances may lead to a state of 'quiescence'. But many insects, particularly in temperate latitudes, periodically enter a state of arrested growth which is not readily eliminated when conditions again become favourable. For this more deep-seated arrest the term 'diapause' is commonly used.

Diapause. The state of diapause may supervene at different stages in different insects: in the recently laid egg, or the egg in which the germ band is newly formed (e.g. in the silkworm and many other Lepidoptera), in the egg containing a fully developed larva (in the mosquito *Aëdes aegypti*, &c.), in one or other of the larval stages, or in the pupa. This state is most commonly an adaptation for survival without feeding during the winter. It often appears to arise without any external stimulus, but that is because the stimulus that induces diapause may occur weeks or months before the effects become visible. The stimuli concerned are closely related to the season: they may be a fall in temperature, or the poor nutritive quality of foliage in the late summer; but the most important of all is generally the length of day.

Photoperiod and diapause. Many Lepidoptera, for example the cabbage butterfly *Pieris*, which normally pass the winter in a pupal diapause, can be induced to breed continuously if the larvae are exposed to a length of day which exceeds fourteen hours or so (Fig. 18, A). In the silkworm, which lays its eggs early in the summer, the effect of day length is reversed: if the female moth is exposed to a day length of fourteen hours or more she lays hibernating eggs.

In certain species, for example the moth *Acronycta rumicis* in Russia, the critical daylength required to induce diapause varies widely in different geographical races in accordance with the latitude to which they are adapted: a given race transplanted to a different latitude may be unable to survive because it fails to go into diapause at the proper time. In some insects, the Nymphalid butterfly *Araschnia levana* is the most familiar example, the diapause pupa gives rise to a 'spring form' which is strikingly different from the 'summer

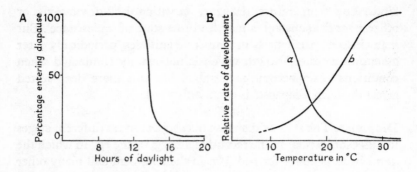

Figure 18. A, diagram showing typical effect of length of day on the occurrence of diapause. Ordinate: percentage of insects entering diapause; abscissa: hours of light in 24 hours.

B, typical example of the influence of temperature on the duration of diapause. Ordinate: relative rate of the developmental process. Abscissa: temperature in °C. *a* 'diapause development', which shows an optimum rate just below 10°C; *b* 'post diapause' development, which shows an optimum rate above 30°C

form' that emerges from non-diapause pupae. Here 'seasonal polymorphism' is genetically linked with diapause.

Termination of diapause. Once diapause is established it is not quickly terminated. Most insects with a winter diapause (such as the silkworm egg) require a period of exposure to low temperature before growth is renewed. It seems that the metabolic processes which have to be gone through before diapause is brought to an end (so-called 'diapause development') have a curiously low optimum temperature. (Fig. 18, B). Certain tropical species, on the other hand, enter diapause in the hot dry season; here the termination of diapause requires a period of exposure to a high temperature.

Hormones and diapause. The immediate cause for diapause is the lack of the hormones necessary for growth and moulting: if the brain from a lepidopterous pupa which is just coming out of diapause after exposure to a period of chilling, is implanted into a pupa that is still in diapause, this pupa will resume its development. Exactly how the seasonal changes in the environment bring about these effects is not known. In the case of the silkworm there is a secretory centre in the suboesophageal ganglion which liberates a 'diapause factor' which in turn influences the developing eggs so that they become hibernating or diapause eggs.

BIBLIOGRAPHY

BODENSTEIN, D. *Insect Physiology* (K. D. Roeder, Ed.), Wiley, New York 1953, 780–821 (control of embryonic development); 822–865 (post embryonic development and differentiation); 866–878 (regeneration); 879–932 (control of moulting and metamorphosis)

CREW, F. A. E. *Sex-determination*, Methuen's Biological Monographs, London, 1954

DANILEVSKII, A. S. *Photoperiodism and Seasonal Development of Insects*, Oliver and Boyd, Edinburgh & London, 1965, pp. 283

DANILEVSKII, A. S. *et al.*, *Ann. Rev. Entom.*, **15**, (1970), 201–244 (day-length and diapause)

GILBERT, L. I. *Physiology of Insects* I (Morris Rockstein, Ed.) Academic Press, New York, 1964, 149–225 (hormones and growth)

KENNEDY, J. S. (Ed.) *Insect Polymorphism*. Symposium No. 1 (1961) Roy. Ent. Soc. Lond.

LAWRENCE, P. A. *Adv. Insect Physiol.*, **7**, (1970), 197–266 (polarity and pattern in insect development: review)

LEES, A. D. *The Physiology of Diapause in Arthropods*, Cambridge University Press, 1955

——*Adv. Insect Physiol.*, **3**, (1966), 207–277 (control of polymorphism in Aphids: review)

WIGGLESWORTH, V. B. *The Physiology of Insect Metamorphosis*, Cambridge University Press, 1954, 152 pp.

—— *Adv. Insect Physiol.*, **2**, (1964), 243–332 (hormones and insect growth: review)

——*The Principles of Insect Physiology*, 7th Edn., Chapman and Hall, London, 1972, 1–26 (development in the egg); 61–145 (moulting, metamorphosis, polymorphism, regeneration, diapause, etc.)

——*Insect Hormones*, Oliver and Boyd, Edinburgh, 1970, 159 pp

WILDE, J. DE, *Ann. Rev. Entom.*, **7**, (1962), 1–26 (photoperiodism in insects: review)

8 Reproduction

Reproduction in most insects is bisexual; the egg cell liberated by the female will develop only after fusion with the spermatozoal cell set free by the male. The physiology of reproduction deals with the arrangements for the separation and ripening of these male and female gametes, and with the mechanisms by which they are brought together. The reproductive system consists of paired sexual glands, the testes of the male and the ovaries of the female, paired gonoducts of mesodermal origin into which the sexual products are discharged, and a median duct lined with cuticle, derived by invagination from the ventral body-wall, forming the vagina in the female and the ejaculatory duct in the male.

Male and Female Reproductive Systems

Testis and spermatozoa. The testis of the male is made up of a series of tubular follicles of varying number and arrangement. Each follicle contains a succession of zones in which the sex cells are in different stages of development. Each 'spermatogonium' arising in the 'germarium' at the apex of the follicle divides repeatedly to form a cyst filled with 'spermatocytes' covered with a mantle of somatic cells. The spermatocytes divide to form 'prespermatids' and then 'spermatids' which, finally, in the lower extremity of the follicle, become transformed into the elongated and flagellated spermatozoa. The spermatozoa are usually discharged through an intromittent penis or 'aedeagus' in adherent bundles or packets mixed with the secretion of a pair of accessory glands. (Fig. 19, A).

114

Ovary and egg cells. The ovary of the female is likewise composed of a series of tubular follicles or 'ovarioles'. In these the oöcytes arising in the germarium do not multiply repeatedly as do the germ cells in the male but each becomes enormously enlarged to form the egg cell or ovum. The oöcytes become enclosed in a mantle of 'follicular cells'; they are nourished, at first by special 'nurse cells', and then by the transfer of materials from the blood by the follicular cells, until they become laden with yolk (Fig. 19, B).

Oögenesis and Oviposition

Yolk formation. In the primitive 'panoistic' type of ovary (Fig. 20, A), as in *Periplaneta*, in which nurse cells are wanting, ribonucleic acid and protein are synthesized by the oöcyte itself through the action of the oöcyte nucleus. All the nutritive materials are supplied by the follicular cells. In the more advanced 'meroistic' types of ovary, as in Lepidoptera and Diptera (Fig. 20, B), where each oöcyte has a group of nurse cells enclosed with it in its follicle ('polytrophic' group), or in Hemiptera, where the nurse cells are confined to the apex of the ovary and are connected to the developing oöcytes by long nutritive cords ('acrotrophic' group), (Fig. 20, C) the nurse cells are mainly responsible for the supply of ribonucleic acid along with glycogen and lipid droplets, which can be seen streaming into the oöcytes in the early stages of their development. Proteins are mostly preformed in the fat body, set free into the haemolymph, and transmitted to the egg by the follicular cells; these cells doubtless play a part also in the synthesis of the lipid and carbohydrate components of the yolk.

Nutrition. The follicular cells then lay down the chorion or egg shell. Finally the follicle ruptures, 'ovulation' occurs and the fully formed egg is discharged into the oviduct and so to the vagina. Arising from the vagina there is usually a duct leading to a sac or 'spermatheca' in which the spermatozoa received from the male are stored. There is sometimes a dilatation or invagination of the vagina (the 'bursa copulatrix'); and there are nearly always 'accessory glands'.

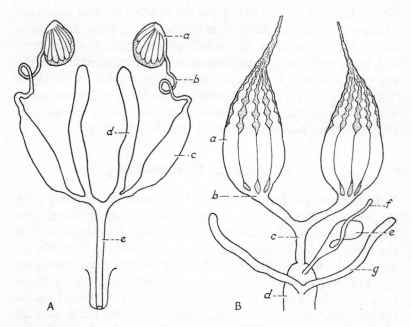

Figure 19. A, male reproductive system, diagrammatic, (after Snodgrass). *a*, testis with follicles discharging into *b*, vas deferens; *c*, vesicula seminalis; *d*, accessory gland; *e*, ejaculatory duct.

B, female reproductive system, diagrammatic (after Snodgrass). *a*, ovary consisting of bundle of ovarioles; *b*, calyx; *c*, common oviduct; *d*, vagina; *e*, spermatheca; *f*, spermathecal gland; *g*, accessory gland

Egg membranes. Besides the chorion laid down by the follicular cells, the egg is enclosed by inner membranes which are deposited by the oöcyte itself. These may amount to no more than a thin layer of wax which serves to waterproof the egg, and a more or less delicate 'vitelline membrane' beneath; or the blastoderm of the developing embryo may lay down a tough chitinous layer, similar to the endocuticle of the free living insect, which often forms the most substantial covering of the egg. These layers may be provided with an elaborate respiratory system which conducts oxygen to the surface of the yolk. In addition, the chorion is generally pierced by

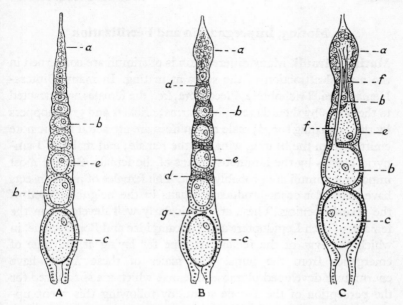

Figure 20. Diagrams showing the chief types of ovarioles (after Weber). A, panoistic type; B, polytrophic type; C, acrotrophic type. *a*, germarium; *b*, oöcytes; *c*, fully formed egg with chorion; *d*, nurse cells of polytrophic ovary; *e*, follicle cells; *f*, nutritive cords of acrotrophic ovary; *g*, degenerating nurse cells

one or more fine ducts, the 'micropyles' through which the spermatoza make their way to fertilize the egg.

Oviposition. The eggs are carried down the oviducts by waves of peristalsis, and deposited singly or in masses according to the habits of the species. They are generally coated with the secretion of the accessory glands which may serve to glue them to the surface or to bind them together in a protective sheath. In the Hydrophilid water-beetles the accessory glands produce silk from which an elaborate 'egg cocoon' is formed; in the cockroach the eggs are cemented together by fluid which hardens to form a tough 'oötheca'. In this insect the protein secreted in the left colleterial gland is tanned and hardened by means of the quinone formed by oxidation of protocatechuic acid secreted in the right colleterial gland.

Mating, Impregnation and Fertilization

Mating stimuli. Many different kinds of stimuli are concerned in the mutual attraction of the sexes at mating. In many Diptera-Nematocera, Trichoptera, Plecoptera, &c., the females are attracted to the males by their dancing in swarms; crickets and grasshoppers by their chirping songs; male mosquitoes are attracted by the note emitted from the beating wings of the female; and males of Lampyrid beetles by the luminous organs of the female. But the most important stimuli are probably scent. The females of many insects have glandular scent-producing organs in the neighbourhood of the sexual opening. These are particularly well developed in the females of such Lepidoptera as Lasiocampidae and Bombycidae, in which the eggs of the female are ripe for laying at the time of emergence from the pupa. The males of these moths have enormously developed plumose antennae which are specialized for the perception of the female scent. By following this scent upwind they can locate the female from a distance of several miles, and are readily attracted to other objects contaminated with the scent. In the silkmoth *Bombyx*, the gipsy moth *Lymantria*, the cockroach *Periplaneta*, and many other insects the chemical structure of the sexual scent has been worked out. In the honey-bee the 'queen substance' secreted by the mandibular glands serves also as a sexual attractant during the nuptial flight. In many Lepidoptera there are scent scales or 'androconia' also in the male; the scent which they produce serves to excite the female to accept the courting male.

Insemination. The transfer of sperm to the female which takes place at mating is often an extremely elaborate process. The aedeagus or penis of the male after entering the vagina may actually penetrate up the duct of the receptaculum so that the spermatozoa mixed with the secretion of the accessory glands are ejaculated directly into the spermatheca. But more often they are discharged into the vagina or into the bursa copulatrix which arises from it.

Spermatophores. In most insects the sperm are not conveyed to the female in a free fluid but are enclosed in a membranous sac or 'spermatophore' formed by the secretion of the male accessory glands. In the primitive Collembola and Thysanura these sperm packets are just deposited at random on little stalks; on coming into contact with the female vulva these spermatophores rupture and the spermatozoa enter the vagina. But in many insects, such as Lepidoptera, crickets, many beetles, the spermatophore is a complex structure which is formed during copulation and transferred with its contents to the female genital tract.

Sperm transfer. In some insects, such as the Gryllidae, the spermatozoa are forced out of the spermatophore by the swelling of the inner walls of the sheath. In any case, whether the sperm are discharged into the female directly or by way of a spermatophore, they are next transferred to the receptaculum seminis or spermatheca. The mechanism of this tranfer is uncertain. Sometimes it is ascribed to pumping or aspiratory movements within the female system; sometimes to active migration by the spermatozoa in response to chemical or other stimuli. In *Rhodnius* the accessory glands of the male secrete an active material (perhaps identical with the material that causes acceleration of the heart beat, p. 37) which acts on nerves to the oviducts of the female and evokes the peristaltic movements that transfer the sperm to the receptacula. A similar secretion occurs in the male accessory glands of many insects. Some idea of the complications is afforded by the fact that the receptacular duct in Lepidoptera has a main lumen up which the sperm pass on their way to the receptaculum, and a very fine subsidiary lumen, or 'fertilization canal', in the substance of the wall along which they descend for fertilization of the eggs.

Fertilization. Fertilization does not take place until long after copulation, usually just before the egg is laid. A few spermatozoa leave the receptaculum and and enter the micropyle of the egg just as it passes the receptacular duct. Very little is known about the mechanism of this process. The egg is usually so orientated that the micropyle is directed towards the duct; and in some insects, notably the honey-bee, there are muscular valves which can limit the number of spermatozoa that escape from the receptaculum. But the factors which control the entry of spermatozoa through the micropyle are very incompletely understood.

Maturation. At the time of ovulation the nucleus of the egg has not yet undergone maturation; but the chromatin has already collected into chromosomes and the first maturation spindle has formed. Maturation seems to be stimulated by the entry of the spermatozoa. Several spermatozoa usually enter the insect egg (polyspermy), but only one of these is normally concerned in fertilization. As it approaches the egg nucleus and, losing its tail, becomes converted into the male pronucleus, the two maturation divisions of the female follow rapidly upon one another. One of these is the reduction division which halves the chromosome number. The two divisions result in the formation of the polar bodies and the female pronucleus which then proceeds to fuse with the male pronucleus to form the zygote.

Factors controlling Fertility and Fecundity

Insects with a short imaginal life copulate once only and lay their eggs in a single batch; others lay batches of eggs with intervals between and may copulate repeatedly; others, again, lay single eggs at fairly regular intervals. In all these groups the egg production or fecundity of the female and the fertility of the male are influenced by a great variety of factors.

Temperature. Like other processes of metabolism, the rate of egg production is markedly influenced by temperature, and usually egg production ceases at low temperatures at which other physiological activities continue normally. At the upper temperature range the male seems to be particularly sensitive: males of *Drosophila* and *Ephestia* become sterile if reared at temperatures only a few degrees above their optimum. In *Rhodnius* reproductive activity in both sexes is completely inhibited at 34°C; this appears to result from the arrest of protein synthesis (cf. p. 104).

Nutrition. Nutrition, of course, has a powerful effect on reproduction, particularly on fecundity in the female. Some insects require water, others require carbohydrate, if they are to produce their full complement of eggs. But the most generally important food factor is protein: flies such as *Calliphora* or *Lucilia* can be kept alive for long periods on sugar and water alone, but the females must have at least one meal of protein before eggs are laid. There are other factors, such as lecithin, which may be essential for some insects; as indeed are vitamins. The fecundity of *Drosophila* females is largely influenced by the quantity and variety of yeast in the diet; and the *Rhodnius* female deprived of its intestinal actinomyces fails to produce eggs (p. 81). Sometimes the nutrition of the larva has an important influence; that is particularly so in those mosquitoes which will produce eggs from reserves accumulated in the larval fat body without feeding at all in the adult stage; and in many insects the abdominal muscles and the flight muscles undergo autolysis in the adult female and provide protein for egg production.

Impregnation. Some insects will produce a normal number of eggs without being mated; but in most the full development of eggs occurs only after impregnation. Sometimes this is an effect merely on oviposition: the eggs are formed but are not laid unless mating has occurred. On the other hand, in many insects the development of eggs is deficient in the absence of the male. In some insects, mating with a sterile male (in which spermatophores or accessory gland secretion are produced) is ineffective; and that has led to the suggestion that the moving sperms themselves provide a tactile stimulus acting through the nervous system. In *Drosophila* and in *Aëdes* a peptide ('matrone') secreted by the accessory glands (paragonia) of the male induces ovulation and oviposition and at the same time suppresses sexual receptivity in the female and prevents further insemination.

Hormones. If the insect is inadequately nourished, particularly if it is deprived of protein, the young oöcytes develop in the follicles until they reach the stage at which yolk should be deposited. Then they die and are absorbed and a new oöcyte moves down. This process of degeneration results from the lack of secretion from the corpus allatum. We saw (p. 102) that the corpus allatum ceases to secrete the juvenile hormone at the final moult, so that metamorphosis can occur. But in the adult insect the same hormone is again produced and is usually necessary for the deposition of yolk. That has been shown in *Rhodnius*, in grasshoppers (where the juvenile hormone is needed also for the secretory activity of the oviducts which form the oötheca), in *Dytiscus* and other beetles, and in Diptera. It seems not to be necessary in some Lepidoptera, such as the silkmoths *Bombyx* and *Hyalophora* which develop their eggs during the pupal stage; but in *Pieris* in which the ovaries mature after emergence the corpora allata are needed for yolk formation. In the male of *Rhodnius* and other insects the corpus allatum secretion is necessary for the activity of the accessory glands which produce the spermatophore.

Like other effects of the corpus allatum, these responses can be evoked by natural or by the synthetic juvenile hormone (p. 104) or by more or less related compounds. The same substance is active in the control of metamorphosis and of yolk formation.

In many insects, notably in the blowfly *Calliphora*, in the cockroach *Leucophaea* and the locusts *Schistocerca* and *Locusta*, the potato beetle *Leptinotarsa* and in mosquitoes, the neurosecretory cells in the dorsum of the brain are even more important than the corpus allatum. Their secretion plays an essential part in stimulating the fat body and other tissues to produce the specific protein required for incorporation in the yolk. The relation between the action of the corpus allatum and of the neurosecretory cells has been much studied but consistent interpretations are still wanting; there is evidence that in some insects both hormones stimulate the synthesis or liberation of specific yolk proteins.

Sensory stimuli from the environment, such as the odour of the normal foodplant for the larva in many Lepidoptera and in locusts, which set in motion the reproductive machinery, probably do so by inducing the appropriate hormone and pheromone secretion in the female. Perhaps the reproductive activity induced in the rabbit flea by the pregnant doe rabbit is of this nature.

Hibernation and Diapause

Reproduction, like growth, may suffer a periodic arrest; and as with growth (p. 111) this arrest may be a direct effect of an adverse environment, or it may be a true 'diapause' which persists even under favourable conditions. In *Dytiscus* and various other beetles, the gonads revert to a resting state after the first reproductive period; they show renewed activity about the same date the following year, sometimes in a third or even a fourth year also. This seems to be a deep-seated rhythm and not a simple effect of warmth following the winter cold. Presumably the process is controlled ultimately by seasonal changes in the environment acting by way of the central nervous system.

In *Leptinotarsa*, and probably in other insects, the arrest of reproduction is induced by the short day-length (photoperiod) in autumn. But the immediate cause of the arrest in these insects is the absence of secretion from the corpus allatum: ripe eggs may be produced in *Dytiscus* at any season of the year if active corpora allata are implanted; and the same applies to *Leptinotarsa*.

Special Modes of Reproduction

Ovoviviparity. The eggs of some insects are retained within the genital tract of the mother until development is well advanced. In *Cimex*, in which impregnation takes the form of the introduction of sperm into the body cavity by way of the so-called 'organ of Ribaga', fertilization takes place in the ovary and the embryo is already developing by the time the egg is laid. In various Tachinid flies the eggs contain fully formed larvae and these escape from the chorion during oviposition (ovoviviparity). In Sarcophagidae and some Oestidae, such as the sheep bot, a smaller number of eggs are present and these may hatch inside the 'uterus', as the enlarged lower part of the common oviduct is called. And in other Diptera, such as *Musca larvipara* or *Mesembrina meridiana*, only one large egg passes at a time to the dilated uterus; it hatches when it is laid. In *Hylemyia strigosa* the larva has already reached the second instar before it leaves the egg.

Viviparity. In none of these cases does the embryo receive any nourishment from the mother during its stay in the uterus. But in the tsetse-fly *Glossina* and in the Pupipara the larva, which hatches in the uterus from an egg of normal size, is nourished until fully grown by special 'milk' glands – modified accessory glands which branch throughout the abdomen. These larvae are matured singly, breathing by the extrusion of their posterior spiracles through the female opening; they are not deposited until they are ready to pupate. In some insects, such as the parthenogenetic forms of Aphids, embryonic development continues in the ovary: the embryo, enclosed in very delicate membranes, is nourished by the follicular epithelium until embryonic development is complete. These are the most familiar of 'viviparous' insects.

Polyembryony. Polyembryony is an unusual mode of asexual multiplication which occurs in some parasitic Hymenoptera. After the maturation of the egg the polar bodies may divide amitotically to form a nutritive sheath; meanwhile the egg nucleus undergoes complete cleavage (for these eggs contain almost no yolk) and the separated cells each develop into embryos. In this way two or four embryos may arise from a single egg in some species; anything from a hundred to two thousand in others.

Paedogenesis. In a few insects there is precocious reproduction in the larval stages. The classic examples occur among Cecidomyidae. The phenomenon, which is known as paedogenesis, was first observed by Wagner in 1861 in larvae of *Miastor*. The oöcytes develop parthenogenetically within the ovarioles until a number of daughter larvae are produced. These are set free into the body-cavity where they rapidly destroy the tissues of their maternal host and ultimately escape through the body-wall.

Hermaphroditism. There are a few insects which are functional hermaphrodites. The best known example is the Californian race of the Coccid *Icerya purchasi*, the fluted scale. The gonads produce both sperm and oöcytes. Most of the eggs are fertilized by sperm from the same individual, but these hermaphrodite females may copulate with the rare males.

Parthenogenesis. Parthenogenesis on the other hand is widespread among insects. Many species, for example many grasshoppers, and some Lepidoptera, can occasionally develop from eggs which escape fertilization. Others, such as the familiar stick insect *Carausius morosus*, are constantly parthenogenetic; males of this species are exceedingly rare. Another classic example is the honeybee, in which the queen can control the escape of sperm from the spermatheca and can thus ensure whether fertilized eggs, which develop into workers or queens, or unfertilized eggs, which become drones, are produced. We have already referred to the familiar example of the Aphids which give rise to a series of parthenogenetic viviparous generations of females throughout the summer months and then, under the influence of the shortened day-length in autumn, undergo a polymorphic change (p. 109) and produce sexual forms, males and oviparous females which lay a small number of fertilized winter eggs.

BIBLIOGRAPHY

BONHAG, P. F. *Ann. Rev. Entom.*, **3**, (1958), 137–160 (ovarian structure and vitellogenesis: review)

ENGELMANN, F. *The Physiology of Insect Reproduction*, Pergamon, Oxford and New York, 1970, 307 pp

JACOBSON, M. *Insect Sex Pheromones*, Academic Press, New York, 1972, 396 pp

TELFER, W. H. *Ann. Rev. Entom.*, **10**, (1965), 161–184 (control of yolk formation: review)

WIGGLESWORTH, V. B. *Adv. Insect Physiol.*, **2**, (1964), 243–332 (hormones and reproduction: review)

——*The Principles of Insect Physiology*, 7th Edn., Chapman and Hall, London, 1972, 700–763 (reproduction, sex determination, etc.)

WILDE, J. DE, *Physiology of Insecta* I (Morris Rockstein, Ed.) Academic Press, New York, 1964, 9–90 (reproduction)

9   Muscles and Movement

Muscle structure and innervation. All the muscles of insects, whether they be 'visceral muscles' surrounding the alimentary canal or the heart, or 'skeletal muscles' actuating the appendages, are made up of striated muscle-fibres. The histology of these fibres varies widely. They may consist of bundles of fibrillae with inconspicuous striations, enclosed in lobulated masses of cytoplasm; fibres like those of vertebrates with highly developed striations and peripheral nuclei; 'tubular muscles' in which the nuclei form an axial core along the fibre; or the peculiar 'fibrillar muscles' which compose the indirect flight muscles of Hymenoptera and Diptera. Each giant 'fibre' in this last type of muscle consists of a bundle of very large fibrils or sarcostyles $2 \cdot 5$–3 μm in diameter, with rows of giant mitochondria or 'sarcosomes' lying between them.

The fine structure of insect muscle as seen with the electron microscope agrees with that in vertebrates. It consists of arrays of 'sliding filaments', stout filaments of myosin and fine filaments of actin, which overlap more deeply when the muscle shortens in contraction. The details of this structure varies widely in different types of muscle.

Where the muscles are attached to the body wall it is usually possible to see fibrillae ('tonofibrillae') passing through the epidermal cells and running into the substance of the cuticle to gain attachment to the epicuticle. In most of the skeletal muscles it is possible to observe a double innervation: one nerve is responsible for rapid muscular contractions, the other for the slow contractions of muscle tonus. In a few insects a third axon with an inhibitory effect is present, which accelerates relaxation.

The Physiological Properties of Insect Muscles

Muscular power. Some insects are able to lift weights of greater mass than their own bodies, and leaping insects can project themselves great distances through the air. But these achievements are a simple result of their body size. For the power of a muscle varies with its cross-section, that is, with the square of one linear dimension; while the volume or mass of the body varies with the cube of the linear dimensions. Consequently, as the body becomes smaller the muscles become relatively more powerful. The absolute power of a muscle is defined by the maximum load it can raise per square centimetre of cross-section. When expressed in this way there is no great difference between the muscles of insects and of vertebrates. Thus the value for man is 6–10 kg per cm^2, for the hind-legs of the long-horned grasshopper *Tettigonia* 4·7 kg.

Properties of insect muscle. When the isolated muscles of insects are studied by the methods of classical physiology they show much the same properties as those of vertebrates: they have a similar 'latent period' between stimulus and response, similar 'summation' when a second stimulus follows closely upon the first, and a similar frequency of stimulation is necessary to induce a steady state of contraction or 'tetanus'. But the results do depend upon whether the 'slow' nerve fibre or the 'fast' fibre has been excited. For example, in the leg muscles of the cockroach, the mechanical fusion of twitches begins at a frequency of 30 stimuli per second when the slow fibre is excited; whereas if the fast fibre is excited it always produces a brief and powerful tetanus, apparently not susceptible to graduated control.

Nervous stimulation. When an insect muscle is stimulated by means of its nerve the conduction of the impulses is carried out solely by the nerve fibres: there is no conduction of impulses along the muscle fibres such as occurs in vertebrates; and if the nerve is allowed to degenerate the muscle can no longer be excited by any form of electrical stimulation. In the wing muscles of *Periplaneta* the contractions begin to fuse at rates of stimulus above 40–50 per second. During flight the wing movements in this same insect occur at a frequency of only 25–30 per second and there are changes in electrical potential occurring at the same rate. It has therefore been concluded that in this insect, in dragon-flies, Lepidoptera, and many other insects there is a nerve stimulus of the conventional type inducing each contraction that leads to a wing beat.

Oscillating fibrillar muscles. But in some of the higher insects, notably in Diptera and in Hymenoptera, the muscular contractions occur at very high rates of 100–200 or more per second. In these cases the muscle contractions are taking place at a far higher rate than the changes in electrical potential. It has been concluded that the relatively infrequent impulses in the nerves serve merely to bring the fibrillar flight muscles of these insects into a reactive state. In this state the fibrillar muscles which control the wings in Diptera and Hymenoptera, or the sound-producing timbals in cicadas, are capable of extremely rapid oscillatory contractions. These muscles commonly operate a so-called 'click' mechanism such that when they have contracted up to a critical point, the thoracic wall or the timbal cuticle clicks over into a new position and the contracting muscles are suddenly released. This sudden release deactivates the contracting muscle and causes it instantly to relax; at the same moment the sudden stretching of the opposing muscle excites it instantly to contract.

Tonic contraction. In additon to their alternate contractions and relaxations the muscles may enter a state of prolonged steady contracture or 'tonus'. In this state they may support the insect in some characteristic attitude and may have to bear a considerable weight; but the rate of metabolism is very low. In the cockroach in the standing position there is a steady discharge of impulses passing down the slow fibres to the depressor muscles of the leg and producing the tonic contraction which raises the animal off the ground. Microscopic examination of a cockroach muscle in this state shows that the contraction is produced by many fibres responding in turn. On this background are superimposed short bursts of impulses in the quick nerve fibres, producing vigorous contractions in the mucles and moving the animal rapidly over the ground. Slow movements are achieved entirely by means of the slow fibres. In soft-bodied insects such as caterpillars, the tense form of the body is maintained by the tonic contraction of the so-called 'turgor muscles' of the body-wall.

Locomotion

Mechanism of walking. As was shown by Johannes Müller a century and a half ago, the insect rests during walking on a supporting triangle, formed by the anterior and posterior limbs on one side and the middle limb on the other, while it carries forward the other three legs. The fore-leg acts as a tractor, the middle leg serves for support, while the hind-leg acts as a propulsor and also turns the body in the horizontal plane. Thus, as the insect walks the centre of gravity falling within the supporting triangle of limbs is carried forwards and outwards towards the apex of the triangle until it falls outside this base and its support is taken over by the other triangle of legs; hence the body zig-zags slightly from right to left as it advances. In *Periplaneta*, running at high speed, the alternating tripod gait is retained; but if a leg is amputated the timing of the leg movements is largely influenced by feed back from peripheral sense organs.

In the walking of caterpillars there is a co-ordination between the 'turgor muscles' which are maintaining the general form of the body and the true 'locomotor muscles' which move the limbs. For the extension of any part of the body is brought about by the relaxation of its own muscles while the general internal tension is maintained by the turgor muscles elsewhere. In other larvae, such as maggots of flies, legs are entirely absent and progression is effected by peristaltic movements or lateral twisting movements of the body-wall combined with friction against the surface due to backwardly directed spines.

Adhesive organs. In walking upon rough surfaces, insects make use of their tarsal claws to maintain their hold; but many can walk or climb on smooth leaves or on glass. For this purpose they use 'adhesive organs' which commonly take the form of expansions of the tarsal surface or of special 'pulvilli'. These may perhaps function in more than one way, but the usual mechanism seems to be the provision of 'tenent hairs', moistened with some greasy secretion, which have soft delicate extremities that can be brought into such intimate contact with the surface that seizure or adhesion takes place and the insect is held by the surface molecular forces.

Leaping. Jumping by many insects is effected by a spring and catch mechanism, the necessary elastic energy being stored in some skeletal structure and the spring being released by a separate muscle. Thus in the flea energy is stored in a pad of resilin; in locusts, in the elastic components of the extensor muscle system.

Flight

Flight muscles. The most characterstic form of locomotion among insects is flight. This is effected chiefly by 'indirect muscles': vertical and longitudinal columns which deform the thoracic capsule by their contraction and so move the wings which are articulated to it. The vertical muscles elevate the wing, while the longitudinal muscles depress it. The complex articulation of the wing gives a wide amplitude to the small but powerful movement of the wing base: in the wasp, for example, the wings vibrate through a sector of 150°. In addition, there are direct muscles inserted into the wing itself. Some of these serve for the flexion and extension of the wings; others will rotate the wing around its long axis or make other adjustments in the wing stroke. In strong fliers like the dragon-fly *Aeschna* the flight muscles comprise 24 per cent. of the total body weight; in the honey-bee they make up 13 per cent.

Wing movements. In the more primitive insects, Orthoptera, Neuroptera, Isoptera, Odonata, the fore- and hind-wings are moved independently during flight, but in Hymenoptera, Lepidoptera, Hemiptera, &c., fore- and hind-wings are linked together to form a functional unit, while in Diptera only the forewings remain. At the base of each wing there is an elaborate articulation which causes the wing to twist in its long axis during its upward and downward movement. This reversal of inclination during the elevation and depression of the wing produces the same mechanical effect as the revolution of a propeller blade. Each wing, in fact, acts as a propeller which draws air from above and in front and drives it backwards in a concentrated stream. The flying insect thus creates a zone of low pressure above and in front, and a zone of high pressure directly behind it.

Aerodynamics and economy of energy. Except perhaps in the very smallest insects, which have wings made of bristles instead of a continuous membrane, the flight of insects is based on conventional aerodynamic principles. The wings act as aerofoils and the variation in lift is controlled by differences in wing twisting, by the frequency of wing beat, and by the total angle over which the wing is moved. It might be thought that the repeated reversal of movement of the wings would be very wasteful of energy; but much of the energy utilized in depressing the wing is in fact stored as elastic energy in the walls of the thorax (in locusts and large moths) or by the elastic components of the wing muscles (in dragon flies) and is released again to help in raising the wing. The flight muscles are thus aided by an elastic oscillation. Elastic energy is stored and released in a slightly different manner by the click mechanism already described (p. 128).

Equilibrium in flight. The insect in flight is faced with the problem of maintaining its equilibrium. In such insects as Tipulids with a long abdomen and long legs, stability may perhaps be maintained mechanically. But many of the best fliers, such as Muscid flies, are inherently unstable. They maintain their equilibrium by the active control of the wing movements. They derive the necessary information from several sense organs such as the eyes, as well as the antennae which perceive air movements. But perhaps the most important sense organs are the halteres. These are modified hindwings richly supplied with campaniform organs (p. 141) which can detect strains set up in the cuticle. Such strains can be set up in the vibrating halteres by the inertia forces exerted when the flying insect deviates from its normal course. The halteres act as gyroscopic sense organs adapted to perceive deviations from their plane of vibration. In the dragon-fly *Anax*, also, the flying insect maintains its equilibrium by means of at least three sensory reactions: a reaction to the general visual pattern, a reaction which ensures that the main source of illumination remains dorsal, and a response to tactile hairs stimulated by the inertia of the head. Furnished with such sense organs good insect fliers can steer accurately; they can hover, go sideways or backwards, or rotate round the head or the tip of the abdomen.

BIBLIOGRAPHY

BOETTIGER, E. G. *Ann. Rev. Entom.*, **5**, (1960), 53–68 (flight muscles and flight mechanism: review)

HUGHES, G. M. *Physiology of Insecta* II (Morris Rockstein, Ed.), 1965, 227–256 (locomotion of insects on land)

PRINGLE, J. W. S. *Insect Flight*, Cambridge University Press, 1957; *Adv. Insect Phsiol.*, **5**, (1968), 163–227 (comparative physiology of the flight motor: review)

SACKTOR, B. *Physiology of Insecta* II (Morris Rockstein, Ed.) 1965, 483–580 (biochemistry of muscular contraction)

USHERWOOD, P. N. R. *Adv. Insect Physiology* **6**, (1969), 205–275 (electrochemistry of insect muscles: review)

WIGGLESWORTH, V. B. *The Principles of Insect Physiology*, 7th Edn., Chapman and Hall, London, 1972, 146–177 (muscles and locomotion in insects)

10 The Nervous System, Sense Organs and Behaviour

The organization of the nervous system. The nervous system is composed essentially of elongated cells which transmit electrical disturbances or impulses from one part of the body to another. These nerve cells or 'neurones' are derived in development from the ectoderm. Each consists of a nucleated cell body and a long filament or 'axon'. Where the bundles of axon filaments run freely through the body cavity they constitute the nerves, which may be afferent (sensory) or efferent (motor). The *sensory neurones* have their cell bodies situated near the periphery. Indeed, they arise throughout post-embryonic development by differentiation from ordinary ectodermal cells which give off an axon process that grows inwards to accompany other sensory nerves on their way to the central nervous system. The cell bodies of the *motor neurones* are situated in the central nervous system, as also are the 'association neurones'.

Nerve impulse transmission. The sensory and motor nerves, with the association neurones, provide the anatomical basis for behaviour. The disturbance or impulse which is propagated along them consists in a change in electrical potential, due to a momentary depolarization of the axon surface, which allows a momentary outflow of potassium ions, passing like a wave throughout the neurone. These waves succeed one another at a rate which varies with the intensity of the stimulus that is being transmitted. The neurones, however, are not continuous with one another. The branched terminations of the axon of one neurone come into intimate as-

sociation with the arborizations of another neurone to form a
'synapse'. It is probable that the electrical disturbance does not
itself cross the synapse but causes the liberation of some chemical
substance, such as acetylcholine, which sets off a fresh disturbance
in the succeeding neurone (see p. 154: 'neurotransmitter
substances').

The simplest type of conduction in the central nervous system will
consist in the transmission of impulses by a sensory neurone from a
receptor or sense organ to the ganglion, through an association
neurone to a motor neurone, and thence to a muscle or other effector
organ. Stimulation of the sense organ will thus produce contraction
in the muscle. This path of nervous conduction is termed a 'reflex
arc' and the response a simple reflex. The reflex arc is a physiological
abstraction; for even in the simplest response the course of conduction must be infinitely more complex, involving inhibition of opposing muscles and compensatory movements elsewhere in the
body. Moreover, the course of reflex conduction is not fixed; as the
synapses become 'fatigued' or 'adapted' the transmission may be
blocked or follow some other reflex arc with a lower threshold; and if
the stimuli are excessively strong they may overflow into many paths
and produce a discharge of impulses from a large group of motor
neurones.

Glial sheaths. Throughout the nervous system the neurones,
both cell bodies and axons, are invested and insulated from one
another by cellular sheaths formed by the glial cells (Schwann cells).
Only at the synapses do the nerve fibres come into intimate contact
with one another. In the central ganglia, modified glial cells named
'perineurium cells' surround the entire nervous system and secrete
over it a tough connective tissue sheath, the 'neural lamella'. Other
glial cells invest the cell bodies, which occupy the peripheral parts
of the ganglia, and the interlacing nerve tracts which form the
central core, or 'neuropile', where the synapses occur. There is no
circulation of haemolymph within the ganglia: the supply of salts
and nutrients and the removal of waste products are effected solely
by translocation through the glial sheaths.

As in vertebrates, the conduction of the nerve impulses along the axons of insects involves the exchange of sodium and potassium between the axon and the surrounding fluid. In many insects the ionic content of the blood is variable and unsuited to this conduction of impulses; but a suitable concentration of sodium and potassium in the fluid immediately around the axon is maintained by the secretory activity of the sheath of glial cells.

The Nerve Cord and Ganglia

The central nervous system of insects generally shows a well-marked segmentation. Typically, each somite has its own pair of nerve centres or ganglia, giving off sensory and motor nerves and connected to the adjacent ganglia by a paired cord containing only nerve-fibres. In virtue of these ganglia, each body segment enjoys a considerable degree of autonomy; probably each has its own respiratory centre controlling the movement of the corresponding spiracles, and each is capable of carrying out such reflex movements as do not involve the activity of other segments. But there is always some degree of fusion between adjacent ganglia; in the higher forms the abdominal ganglia are fewer; in higher Diptera the thoracic ganglia become united into one; in Hemiptera all the thoracic and abdominal ganglia have coalesced to form a single mass.

When a given movement of the body demands the co-operation of a number of body segments, there exists in one of the ganglia a higher centre which co-ordinates the activities of the others. In those insects in which most movements involve only the thorax, as in the winged forms, the ganglia have become enlarged and concentrated, and the isolated thorax is then capable of performing many of the motor functions – walking, clasping, vibration of the wings – of which the whole animal is capable.

The Brain and Suboesophageal Ganglion

Suboesophageal ganglion. There is also a concentration of ganglia in the head: the suboesophageal ganglion and the supraoesophageal ganglion or 'brain'. Of these, the suboesophageal seems to be the equivalent of the motor ganglia that we have been discussing; it controls the movements of the mouth appendages, and may contain co-ordinating centres – as in caterpillars, where it controls the forward walking movements, and in the adult of *Aeschna*, where it inhibits the tenacious clasping reflex of the legs.

Brain. The brain, on the other hand, seems never to contain motor centres of this kind. It seems rather to be the central receptor for the abundant sense organs located in the head; and as such it orders the movements of the whole insect in accordance with the stimuli which these receive. The reflex movements of the isolated body segments are themselves purposive in a limited degree; but when co-ordinated by the brain, all the body movements serve the purpose of the animal as a whole. The 'mushroom bodies' in the brain are collections of association neurones which are regarded as the most important centres regulating behaviour; they are highly developed in ants and bees.

Functions of the brain. If the brain is removed, the reflex reactions of the body become exaggereated; it reacts to stimuli which in the normal insect would have no effect; and movements, such as the cleaning movements in the bee, may continue without cessation for hours at a stretch. Thus one function of the brain, one method by which it disciplines the organism, is to inhibit the too rigid execution of reflexes by the lower centres.

If one half of the brain is destroyed, the insect tends to move in a circle towards the sound side. This reaction is due partly to the increased propulsive activity in the limbs on the operated side following the removal of inhibition; partly to an effect of the brain in increasing the muscular tone of the flexor muscles on its own side. But neither of these mechanisms will provide a complete explanation of the circus movement; for after mutilation of the limbs on either side, the performance of the muscles is so modified that the same movement still occurs. The cause of this behaviour therefore lies deeper: each half of the brain, when acting alone, tends to bring about movement towards its own side – movement that requires the co-operation of muscles in all parts of the body.

Reflex Responses in the Normal Insect

Reflex behaviour. Comparatively simple reflexes undoubtedly play a large part in the responses of the normal insect; and our main problem is to describe the whole of insect behaviour in terms of such reflexes. Pressure upon the tarsus causes a reflex withdrawal of the leg; removal of the feet from contact with the ground evokes a turning over reflex or initiates the vibration of the wings; contamination of the antenna induces cleaning movements. The mechanical character of such reflexes is well seen in the cockroach: if the antenna is stimulated, and the insect is offered a bristle, it will clean the bristle and neglect the antenna; and many insects will continue the cleaning movements after their antennae have been amputated. But though mechanical, these reflexes are purposive and plastic: the water beetle *Dytiscus* deprived of one hind-leg will swim with the middle leg instead; and if the forelegs of the cockroach are removed, the stimulated antenna will be held by the middle pair.

Reflex inhibition. Even the simplest of these reflexes involves many parts of the body, and some movements of the whole insect are also clearly reflex. This is notably the case with certain inhibitions. We have seen how loss of contact between the feet and the ground will initiate flight; conversely, flight is inhibited when contact is regained. And contact with other parts of the body may similarly inhibit movement: the apparent preference for dark crevices in such insects as the moth *Amphipyra*, the earwig *Forficula*, and the bed-bug *Cimex*, depends in fact upon a reflex inhibition of their movements by contact ('thigmotaxis' or 'stereokinesis'). In the state of immobilization or 'akinesis' which such conditions induce, these insects may require strong stimuli to arouse them. Experimentally, a similar hypnotic state may be induced, in the earwig for example, by many kinds of constantly repeated stimulation; and this state is doubtless related to the death-feigning reflex or 'thanatosis' shown by many insects when alarmed.

Giant axons. Insects which are capable of very quick 'escape reactions', such as the cockroach or the larvae of dragon-flies, possess 'giant axons'. These large nerve fibres conduct impulses at a rapid rate; they carry stimuli from sense organs in the abdomen and enable the insect to give a very quick sterotyped response with about the same speed as the eye-blink in man.

Orientation

A great part of behaviour consists in orientation – the direction of the movements of the animal in response to external stimuli. For the purposes of physiological description such reactions are regarded as made up of a series of reflex responses organized in such a way that the animal is moved in a predictable direction. But it is seldom possible to trace in detail the reflexes involved: the most that can be done is to classify the general mechanisms by which certain stimuli can bring about appropriate movements.

Kineses. In the first place the insect may reach its destination without being truly orientated: the direction of movement is not clearly related to the source of stimulus. Reactions of this type are termed 'kineses' or undirected reactions. The simplest form is 'orthokinesis': when the stimulus acts the insect moves, when it ceases to act the insect comes to rest. We have already considered some examples of 'akinesis' brought about in this way.

Klinokinesis. A more complex form of undirected response is termed 'klinokinesis'. Here the insect moves in a straight line in a favourable environment, but as soon as it enters a mildly unfavourable environment it begins to make turns, the frequency of turning increasing with the strength of the stimulus. If the stimulus is very strong the insect turns aside instantly, giving an 'avoiding reaction'. After a time it becomes 'adapted' to the adverse stimulus; it then goes straight, and continues to do so as long as the stimulus remains the same or diminishes. If the adverse stimulus increases, it once more begins to turn. These two factors, of random turning and adaptation, will lead the insect to a favourable environment. Klinokinesis is a mechanism that is used by many insects for orientating themselves in diffuse types of stimuli without steep gradients, such as temperature, smell, humidity or the texture of a surface.

Taxes: klinotaxis and tropotaxis. Where a more complex system of sense organs is present the insect can make use of directed reactions or 'taxes'. Sometimes the insect will compare the intensity of stimulus on the two sides, either by moving the whole body or by swinging the antennae first to one side and then the other. It then moves in the preferred direction. A striking example of this is seen in the maggots of flies which retreat from the light. As they move they swing the anterior end of the body where the light-sensitive organs are, to right and left; they then proceed in the direction in which the organs are most shaded. This procedure of making successive comparisons is termed 'klinotaxis'.

Sometimes the insect may be able to compare the intensity of stimulus on the two sides simultaneously, without movement of the sense organs or of the body; this less common mechanism is termed 'tropotaxis'. It can be strikingly illustrated in the honey-bee. The bee can be trained always to turn towards the scent of aniseed when it is exposed to this scent in a Y-shaped osmometer. But if the antennae are crossed and fixed down in this position, the trained bee always turns in the wrong direction. The term 'anemotaxis' is used to describe the movement of insects in response to an air stream: commonly the insect is excited by the odour in the air stream and moves up or down wind, often using tactile or visual cues.

Telotaxis: a fixation reflex. In the case of vision an object or a point of light is perceived from a distance. The image of such an object may be 'fixed' upon the central point of the retina of the eye and the insect thus enabled to go straight to it. This response is termed 'telotaxis' or movement towards a definite goal. The response seems to be made up of a series of reflexes specifically related to the region of the eye that is stimulated (like the fixation reflex in man); so that the insect finally turns until the image of the source of stimulus is received upon the region of clearest vision (in the insect with a single eye) or upon corresponding points in the two retinae (in insects with both eyes intact). So exact has this reflex become in such insects as the fire-fly *Photinus*, that the single flash of light emitted by the female will cause the male to turn accurately in her direction.

This same 'fixation reflex' doubtless provides the basis of the response to moving objects that is shown by many insects. Some forms, like the larva of the dragon-fly, will follow any object moving near them; others, like the adult dragon-fly, after turning momentarily to such an object, instantly leave it if it is not suitable for food, the initial reflex being inhibited by the higher centres. In the bug *Rhodnius*, the response to moving objects is normally elicited with difficulty, but it becomes very marked if the antennae are removed.

Menotaxis and light compass response. Then there are insects which will move at a constant angle to a source of light; a method of orientation known as 'menotaxis'. In some cases this response seems to result from the balanced turning effect of two lights – that from the source itself and that reflected from the background. Sometimes included in this mechanism is the 'light-compass' orientation of ants and bees, which make use of the direction of the sun's rays in finding their way back to the nest. Certain aquatic insects (e.g. *Notonecta*) maintain their position in running water by a comparable mechanism – by keeping constant the image of objects on the bank. If these objects are moved, the insect can be caused to move with them. Even such insects as Gyrinidae, which swim in all directions in the water surface, seem, surprisingly enough, to maintain their position by the same mechanism.

Sense Organs: Mechanical Senses

Trichoid sensilla. We must now consider briefly the sense organs of insects from the point of view of their relation to these forms of orientation. Most of these sense organs take the form of 'sensilla', small organs in the integument each of which is furnished with a sense cell (or group of sense cells) with a sensory axon. The primary type of sensillum is the tactile hair. Indeed, nearly all the articulated hairs or spines of insects are provided with a sense cell which is stimulated when the hair is moved. Such organs are often most abundant on the antennae, which are thus important tactile organs, but they occur all over the body and the feet and are responsible for the responses of 'akinesis' or 'stereokinesis' already described. The ultimate receptor appears to be an assembly of microtubules within the base of the cuticular hair which are compressed when the hair is moved.

Campaniform and chordotonal sensilla. In some parts of the body sensilla are developed without any spine or bristle: there is just an innervated dome set in the cuticle. These structures are termed 'campaniform sensilla'. They are stimulated by pressure or by the bending of the cuticle acting upon the bundle of microtubules. And, finally, all outward sign of sensilla may disappear; the component parts become elongated and deeply sunk within the body. These structures are termed 'chordotonal sensilla' or 'scolopidia'. They usually lie in the axis of an elastic strand stretched between two points in the body-wall, one of which is some pliable region of the cuticle, such as the intersegmental membrane. They are doubtless stimulated by changes in tension between the two points of attachment.

Proprioception

Postural hairs and stretch receptors. Any of the mechanical sense organs may be stimulated by contact or pressure from the environment. But they may also be stimulated by happenings within the body itself. Rows of spines or patches of hairs may serve to indicate the degree of flexion in a joint; they are called 'postural hairs'. Hair-cushions of this type between the head and thorax, and between the thorax and abdomen, are the chief gravity sense organs in the honey-bee. Chordotonal organs may provide information about the pressure or tension exerted by the muscles in any part of the body. The campaniform organs, by reporting the strains set up in the cuticle, are probably among the most important of these 'proprioceptive organs'. Probably all co-operate in detecting the distribution of stresses in the limbs, &c., brought about by gravity, and thus they serve as organs of equilibrium and enable insects to orientate their movements against gravity (negative geotaxis).

Many insects possess 'stretch receptors' which consist of connective tissue strands with associated nerve cells; they are probably involved in the rhythmical movements of the abdomen and thorax, as in respiration. Over-stimulation of stretch receptors in the foregut and body wall leads to the arrest of feeding; if the nerve cord is cut, feeding will continue until the body wall is ruptured.

Specialized receptors. There are some special developments of these proprioceptive organs which may be noted. There is always a group of chordotonal organs at the apex of the pedicel of the antenna (Johnston's organ) which serves both to detect passive movements of the antenna due to air currents and to estimate the position of the antenna brought about by the action of the muscles. There is often another group in the tibia, which forms a 'subgenual organ' that is sensitive to vibrations of the ground on which the insect rests. And finally there is a remarkable development of all these organs, but particularly of the campaniform organs, in the halteres or reduced hind-wings of Diptera. These structures are exceedingly sensitive to strains set up in the cuticle and, as we have seen (p. 131), they enable the insect to respond to the inertia forces set up in the vibrating haltere when the insect changes direction during flight.

Hearing

Trichoid sensilla and hearing. Where these mechanical sense organs are sufficiently sensitive and suitably disposed they may be stimulated by the pressure changes or the air movements associated with air-borne sounds. That applies to the antennae of mosquitoes in which Johnston's organ is highly developed; the entire bushy antenna of the male mosquito is set in vibration by the high-pitched sound emitted by the female in flight. It is this note which provides the stimulus to mating. Even the cerci of crickets, which are primarily tactile organs sensitive to air movements, may have their long tactile bristles set in vibration by low-pitched sounds.

Specialized organs of hearing. The most highly developed auditory organs, which occur in Tettigoniids, Locustids, Lepidoptera, &c., consist of batteries of specially modified scolopidia stimulated by the movements of a membranous ear-drum or 'tympanum'. In these insects the auditory organs become much more sensitive to sounds of higher frequencies, and their range of perception extends into frequencies far beyond those which the human ear can detect. They possess remarkable powers of recognizing the sounds produced by their own species. They apparently do this by responding, not to the pitch of the sound, which is the characteristic most noticeable to the human ear, but to the frequency with which the note is 'modulated'.

Perception of frequency modulation. Insect sounds are generally produced by a rod (connected with a sound-producing membrane) being moved across a series of teeth; each tooth strengthens or modulates the note given out by the membrane. The tympanal organs of insects have a very small 'time constant': they can distinguish sounds separated by intervals as short as 1/100 sec., whereas the human ear cannot perceive intervals of even 1/10 sec. The insect ear is therefore well adapted to perceive the modulation of sound frequency in the songs of their own species, and Noctuid moths are able to perceive and recognize the rapid pulses of high pitched sounds given out by bats hunting by echo-location and to determine the direction from which they come.

Stimulatory organs. All the mechanical sense organs, besides the specific effects which stimulation of them elicits, have the property of arousing the nervous system generally, and bringing it into a reactive condition. They belong to the so-called 'stimulatory organs' which serve both to maintain the muscle tone, and to enhance the kinetic efficiency of the muscles when they are brought into action. Prominent among the stimulatory organs are the halteres of Diptera and Strepsiptera, one important function of which is to sustain the muscular vigour of the insect during flight by the stimuli which their own vibration generates. In this case the receptors seem to be partly the chordotonal organs but chiefly the abundant campaniform sensilla.

Chemical Senses

Taste and smell. The chemical senses of taste and smell are developed in varying degree. Accepting the definition that what we perceive by the tongue are tastes and the flavours we detect by the nose are odours, insects can recognize the tastes of salt, sweet, acid, and bitter by means of receptors in the mouth (*Apis*, &c.), on the antennae (Hymenoptera), on the palpi (*Dytiscus*, *Geotrupes*, &c.) and on the tarsi (Diptera, Lepidoptera) – often in several of these sites in the same insect. They can detect odours chiefly by sensilla on the antennae, but to a lesser degree by the palps. While the 'general chemical sense', affected by pungent vapours, is present in many parts of the body. Receptors for the perception of specific flavours occur in the pharynx of sucking insects, notably the Hemiptera; thus the blood-sucking bug *Rhodnius* will take up salt solution through a membrane more readily if a trace of haemoglobin is added to it.

Chemoreceptors. The chemoreceptive sensilla are usually thin-walled hairs or pegs. The distal nerve processes from the sense cells break up inside the cavity of the hair to produce a great number of fine filaments, visible only with the electron microscope, which penetrate the thin walls of the sensillum and come to have their extremities freely exposed to the atmosphere. In some of the larger sensory hairs, such as those on the tarsi or labella of blow flies, there are three or four sense cells giving off separate distal processes which serve different purposes – to perceive sugars, salts, repellant substances, or water.

Sensory thresholds. The sensitivity of the organs of taste in insects may be very high; under conditions of starvation, the butterfly *Pyrameis*, and the fly *Calliphora*, may be able to detect with their tarsi concentrations of sugar of $\frac{1}{12,800}$ M or less: the threshold for the human tongue lying between $\frac{1}{64}$ and $\frac{1}{32}$ M. The apparent sweetness of the various sugars may be very different for insects and man; and insects may differ in this respect not only between themselves, but even when the organs of taste on the tarsi are compared with those on the mouthparts: in *Calliphora*, lactose is apparently tasteless to the tarsi, but evokes a response from the mouthparts.

Smell and behaviour. Insects vary enormously in the extent to which their behaviour is directed by the sense of smell. Many species find their food largely by this means (cockroaches, crickets, *Drosphila*, &c.); butterflies often appear to be led to the food plant upon which to lay their eggs by some single odorous constituent in the leaves; parasitic Hymenoptera find by smell the larva in which to lay their eggs, sometimes detecting this from a distance, sometimes following the trail of scent which it has left behind it; the following of trails of scent is very important also in the orientation of ants and termites; but the highest development of this sense is seen in the males of those Lepidoptera (Lasiocampidae, &c.) which can detect and locate the female from distances of more than a mile by means of the olfactory organs in the antennae. Many of the chemoreceptors of insects are made up of cells which are specialized to respond to simple substances, such as specific sex attractants, specific foods, such as carrion in burying beetles, or special components in the food plant, such as mustard oil glucosides in *Pieris* larvae.

The potentialities of the sense of smell have been most fully studied in the case of the bee; and in this respect the bee's perceptions seem to be surprisingly like our own. Flowers odourless to us are odourless to them; substances of unlike chemical composition which smell alike to us – such as nitrobenzol and oil of bitter almonds – are confused also by the bee; and the threshold concentration for odours to be detected seems to be of the same order of magnitude for bee and man. Bees can readily be trained to associate the presence of a supply of food with particular scents; and they can retain the memory of such associations for several weeks – far longer than they can visual memories. But it is interesting to note that they can be trained only with aromatic odours such as might occur in flowers, and not to such substances as indol or asafoetida.

Temperature and Humidity Senses

All insects react to changes in temperature and humidity, but the sense organs responsible are not well defined. The temperature sense is spread all over the surface of the body; but in blood-sucking insects the antennae are particularly sensitive and are used for locating warm-blooded hosts. This response seems always to be to the temperature of the air and not to radiant heat. The sensilla concerned are probably the fine hairs which are abundant on the antennae.

Fine tufts of hair connected with sense cells serve as organs for detecting changes in atmospheric humidity in the louse *Pediculus* and in the larva of *Musca*. But in most insects it has not proved possible to distinguish between thin-walled chemoreceptors and the organs sensitive to humidity changes. Caterpillars can get information about the turgidity of a leaf by means of humidity and temperature receptors on the antennae.

Vision

We have already seen how varied are the visual responses of insects. The receptor organs which serve these responses range from small groups of light-sensitive cells beneath lens-like thickenings of the cuticle, to the compound eyes of the highest forms, each with many thousands of retinal elements and capable of tolerably good image reception.

Stemmata. The stemmata, or simple eyes of larvae, which occur in the absence of compound eyes, are responsible for the orientating responses to light in maggots and caterpillars. In their highest development, which is attained in the predaceous larvae of Cicindelids and other beetles, they serve to locate the prey; and although it is doubtful if they are capable of image perception, they can certainly perceive the movement of objects within the field of view.

Ocelli. The ocelli, or simple eyes of adult insects, on the other hand, are optically unsuited to even the crudest type of image formation;

for not only are the retinal elements few in number, but any image that is formed by the lens must fall far beyond the level of what elements exist. In adult insects it is the compound eyes which are responsible for the movements towards light: such movements do not occur if the compound eyes are covered, and only the ocelli remain.

What, then, is the function of ocelli ? So far, the only suggestion of the many that have been put forward, which has any experimental evidence to support it, is that they are stimulatory organs (p. 143) which accelerate phototaxis by increasing the sensitivity of the brain to light-stimuli received through the compound eyes. Thus the response of *Drosophila* to light is appreciably more rapid and more persistent if the ocelli are intact than if they are blackened. It must be realized that the light sense organs have two functions; a phototactic or orientating function, such as we have already discussed, and a photokinetic or stimulatory function. The phototactic response in *Drosophila* is essentially an escape reaction, manifested only when the insect is alarmed. The photokinetic ocelli increase its efficiency. Honey-bees with their ocelli intact start to forage earlier in the morning and continue later in the evening than bees with the ocelli covered.

Compound eyes. But whenever a faceted compound eye is present, no matter how few may be the retinal elements it contains, these will inevitably receive different intensities of light from different parts of the visual field; and some kind of visual pattern must always result (Johannes Müller's 'mosaic theory' of insect vision). Consequently, any movement in the visual field will at once be perceived by the changes in pattern it will produce.

We have seen that many insects will turn towards an object that is moving; will turn, that is, until corresponding points of the two retinae are stimulated by it. Now the closer the object is to the insect, the nearer to the mid-line will be these corresponding points. Consequently, the insect may enjoy some of the advantages of binocular vision, and be able to judge accurately the distance of objects, even though its perception of images be rudimentary; and like the larva of *Aeschna* or the bug *Rhodnius* (when deprived of its antennae), it may attack with precision any object that moves within range.

Visual acuity. In forms like the adult dragon-fly and the bee, image perception reaches a higher plane, and specific objects can be recognized; so that Sphingid moths, for instance, will endeavour to extract nectar from the flowers of wallpapers or from models of flowers. By calculation and by experiment it has been estimated that the visual acuity of the bee, in the vertical axis, is about one-fiftieth that of man (in the horizontal axis the acuity is only about one-third of this because the eye is more curved and each ommatidium has to cover a wider field – the bee, in fact, is astigmatic). But this degree of acuity is obtained only under optimal conditions of illumination. With poor illumination, both the visual acuity and the discrimination of different light intensities are so greatly inferior to that of man, that the resolving power of the eye must be extremely feeble. None the less, these insects can recognize different forms, and can learn to associate with these the presence of food. The bee and other Hymenoptera can find their way over miles of country by means of landmarks, and can recognize the entrance to their own hive or nest by visual memory.

Apposition and superposition image. According to Exner's elaboration of the mosaic theory there are some insect eyes (such as those of Orthoptera, Hymenoptera, Diptera) in which the light received in one facet illuminates only a single retinal element to produce an 'apposition image'. Whereas in other insects, notably in nocturnal Lepidoptera in the dark-adapted state, there is a wide gap between the inner limit of the cones and the retinal layer. This gap is bridged by refractile filaments which serve as 'light guides' and, by internal reflection of the rays, tend to prevent the dispersion of light to neighbouring ommatidia. But there is some scattering of light, and this is best seen in the skipper butterflies (Hesperidae) in which light guides are absent and light illuminating a single receptor element will enter the eye through quite an extensive group of facets. The resulting image was termed by Exner a 'superposition image'; it will be brighter but less well defined than the 'apposition image'.

In *Musca* a remarkable type of 'neural superposition' exists, in which light stimuli, from a single point in the field of vision, are received by six widely separated retinula cells and are then recombined by the neural route at a single point in the optic ganglia. This elaborate system, applied over the entire eye, will give a degree of resolution that is very much finer than what would be possible under the mosaic theory of Müller and Exner.

Perception of polarized light. Another property which is characteristic of the insect eye is its ability to detect the plane of polarization of light. It might be thought that this ability would be of no great practical importance; but in fact the light coming from a blue sky shows a characteristic pattern of polarization which is related to the position of the sun. Bees are able to orientate themselves by means of this pattern, provided they can see part of the blue sky, even if the sun is concealed. Indeed, it has been found that many insects pursue a perfectly straight course when allowed to walk in the open under a clear sky. It was long believed that this was a sun compass reaction, an example of 'menotaxis' (p. 140) related to the position of the sun. But it appears more often to be a menotactic orientation to the pattern of polarized light in the sky. This ability to recognize the plane of polarization resides both in the compound eyes and in simple eyes, such as the stemmata of caterpillars. The physical mechanism involved is not fully understood; but it seems likely that the retinal rod, or 'rhabdom', contains dichroic visual pigments orientated in a regular fashion characteristic of each radial segment, or 'rhabdomere'. That is, that the physical mechanism is like that of polaroid.

In normal vision the light reaching the rhabdom acts upon the retinal pigment 'retinene' (a derivative of vitamin A) which the rhabdom contains, and the resulting breakdown products stimulate the sensory cells of the retina.

Colour Vision

Visual pigments. The visual perceptions, notably in flower-visiting insects, are greatly enhanced if the objects are coloured;

insects possess, in fact, a well-developed colour vision. The histological basis of this colour vision is not certainly known, but there is evidence that the different rhabdomeres of a single rhabdom may contain different visual pigments and react to different wavelengths of light, so that a single retinal element, below a single facet, will be able to distinguish a range of colours.

Spectral luminosity. The functional significance of colours in the vision of insects is much better understood. Certain responses, such as the fixation reflex, are influenced only by the *luminosity* of the source of stimulus and not by its colour. Such responses, therefore, afford a method of testing the subjective luminosity of different colours for different insects; and it has been shown that for many insects (*Mantis, Coccinella, Apis*) the luminosity of the visible spectrum is the same as for man with the eye dark adapted; for others (*Pieris*) it corresponds with the light adapted human eye. Many insects can vary the amount of light entering the eye, by movements of the pigment in the iris cells. These changes, combined perhaps with biochemical changes in the retina, can bring about a 1000-fold increase or decrease in the sensitivity of the eye.

Colour vision. Corresponding with these differences in apparent luminosity, there are differences in the visibility of the spectrum. Certain insects (*Apis, Macroglossa*, Lep.) are practically blind to red, but can see far into the ultra-violet. Such insects may be attracted by the ultra-violet light reflected from certain flowers; and they will cluster round screens illuminated with ultra-violet rays which are invisible to us. Indeed, the stimulating efficiency of the ultra-violet part of the spectrum, for the honey-bee and for *Drosophila*, may be far greater than that of the regions visible to man; and although this effect may be due in some degree to the fluorescence of parts of the eye under ultra-violet illumination, this cannot account entirely for the phenomenon. There must be a true perception of ultra-violet; and it is interesting to note that the limit of this perception is, approximately, the lower limit of the solar spectrum after the far ultra-violet has been filtered out by the atmosphere. Other insects are capable of seeing red (Pieridae), and associated with this is the shift in the apparent luminosity of colours that we have already noted.

Simultaneous contrast. Insects, like man, experience the phenomenon of simultaneous contrast; a grey field surrounded by yellow appears blue to the bee, a grey field surrounded by blue appears yellow – but in view of the fact that ultra-violet is perceived by the bee as a separate colour, not yellow and blue, but yellow and 'bee-violet', blue and 'bee-purple', and blue-green and ultra-violet are complementary colours. The capacity for colour vision may vary markedly in different parts of the eye.

Insect Behaviour

Perceptions. Most of the information on colour vision in insects, and a great deal of the more reliable information upon other senses, has been obtained by training experiments: stimuli which the insect can learn to associate with the presence of food, or with the location of its nest, are judged to be perceptible. Such experiments, of course, need the greatest care in their interpretation; for the insect may be guided by some other stimulus that has been overlooked by the experimenter; indeed, in the course of a single experiment, it may cease to be orientated by one stimulus and come to depend upon another; and there are many who hold that the whole perceptual experience of the organism is integrated, as it were, into a pattern, and that it responds to general changes in this pattern, and *not* to the isolated stimuli of which it is composed (Gestalt theory).

Inborn patterns of behaviour. Thus, when we consider the behaviour of insects under natural conditions, we find them gaining their ends by an infinite variety of sensory impressions. Inborn reflexes, and reflexes 'conditioned' by the experience of the individual, certainly occur; but these are masked by inhibitions, and integrated by higher centres, in such a way that they serve constantly the needs and purposes of the insect as a whole. It is this unifying quality, that welds the organism together, and makes the whole something greater than the sum of the parts, which in the sphere of behaviour as in the sphere of growth still eludes physiological analysis. In the absence of a satisfactory physiological theory, many of the activities of insects are described in terms of inborn 'patterns of behaviour' which are called 'instincts'.

Inborn patterns exist at all levels in the organism. The pattern of growth is the outcome of the pattern of genes in the chromosomes. The pattern of growth varies with time. The immediate cause of this variation lies in the cycles of hormone secretion; and these cycles may be correlated with changed patterns of behaviour.

Diel cycles. Other cyclical changes, both in metabolism and in behaviour, are linked to the twenty-four hour cycle of the terrestrial day. The twenty-four hour (or approximately twenty-four hour) duration of these diurnal, or circa-dian, cycles is inborn, and they will often continue for some days in completely constant surroundings. But the initiation of a cycle is controlled by cyclical changes in the environment, usually changes in illumination. In the circa-dian cycle of activity in *Periplaneta*, the immediate factor which excites activity is a hormone circulating in the blood. This hormone is the product of neurosecretory cells in the suboesophageal ganglion, or perhaps in the dorsum of the brain; and the timing mechanism, or 'physiological clock', appears to reside in these cells. The nervous system, hormones, growth and behaviour are thus closely interwoven.

BIBLIOGRAPHY

BURKHARDT, D. *Adv. Ins. Physiol.*, **2**, (1965), 131–173 (insect colour vision)

DETHIER, V. G. *The Physiology on Insect Senses*, Methuen, London, 1963, 255 pp.

GOLDSMITH, T. H. *Physiology of Insecta* I (Morris Rockstein, Ed.), Academic Press, New York, 1964, 397–462 (insect vision: review)

HARKER, J. E. *The Physiology of Diurnal Rhythms*, Cambridge University Press, 1964, 114 pp.

HASKELL, P. T. *Insect Sounds*, Witherby, London, 1961, 189 pp.

HUBER, F. *Physiology of Insecta* II (Morris Rockstein, Ed.), 1964, 333–406 (integration by the central nervous system in insects)

LINDAUER, M. *Communication among Social Bees*, Harvard University Press, 1961, 143 pp.

RUCK, P. *Ann. Rev. Entom.*, **9**, (1964), 83–102 (structure of insect retina: review)

SCHWARTZKOPFF, J. *Physiology of Insecta* I (Morris Rockstein, Ed.), 1964, 509–561 (mechanical sense organs: review)

SLIFER, E. H. *Ann. Rev. Entom.*, **15**, (1970), 121–142 (structure of insect chemoreceptors: review)

THORPE, W. H. *Learning and Instinct in Animals*, Methuen, London, 1962

WIGGLESWORTH, V. B. *The Principles of Insect Physiology* 7th Edn., Chapman and Hall, London, 1972, 178–214 (nervous system); 215–255 (vision); 256–309 (mechanical and chemical sense organs); 310–356 (physiological mechanisms of behaviour)

11 The Endocrine System

Throughout this book we have had frequent occastion to mention the role of hormones. Circulating hormones are involved in the operation of all systems of the body. The purpose of this chapter is to bring this information together and to view the system of internal secretions, the endocrine system, as a whole.

Nervous System and Hormones

Trophic nerves. In all nerve cells the products of the cell body, including the mitochondria, move slowly down the axon towards its termination. It is generally supposed that this movement serves trophic, or nutritional, functions and is needed to maintain the axon in a healthy state.

Neurotransmitters. The axon ends either in a synapse or multiple synapses in contact with the processes of other nerve cells, or on the cell body of a muscle or gland. The nerves convey electrical disturbances produced by progressing cycles of depolarization of the axon membrane. At the synapse, including the neuromuscular junction, however, this electrical transmission is interrupted; a chemical substance, termed a 'neurotransmitter' is produced and this serves to set up a progressing cycle of depolarization in the adjacent nerve or in the muscle. In the central nervous system the chief chemical transmitter set free at the synapse is acetylcholine. γ-Aminobutyric acid (GABA) is used as an inhibitory transmitter. Noradrenaline and dopamine also probably play some rôle as neurotransmitters. At the neuromuscular junctions of insects (unlike vertebrates) acetylcholine appears not to be involved; glutamate probably serves as the neuromuscular transmitter.

Various glands exposed to the action of 5-hydroxytryptamine or a similar material circulating in the blood may be stimulated to synthesize adenosine 3' 5'-monophosphate (cyclic AMP) which in turn acts as a chemical messenger within the cell and provokes its secretory activity. That may happen in secretion by the salivary glands, or in the action of a diuretic hormone on the Malpighian tubules.

Neurohumours. Noradrenaline, 5-hydroxytryptamine, acetylcholine and other neurotransmitters may sometimes be liberated into the circulating blood and so have a widespread effect throughout the body. They are then often called 'neurohumours' or 'neurohormones'. But there is no very sharp distinction between these 'neurohumours' and the 'neurosecretions'.

Neurosecretion. Certain cells within the central nervous system contain material which stains conspicuously with certain specific stains, and has a characteristic appearance under the electron microscope, being made up of electron dense spheres 100–400 nm in diameter. This material also is carried down the axons. Such cells are called neurosecretory cells. Those in the dorsum of the brain which produce the activation hormone to initiate moulting (p. 99) and which have an important influence on reproduction (p. 122) are the best known example.

But neurosecretory cells are widespread in other parts of the brain, in the corpus cardiacum, the suboesophageal ganglion and throughout the ganglia of the nerve cord. Some of the examples referred to in this book are the diapause hormone in the silkworm (p. 113); the colour change hormones in stick insects and in *Pieris* pupae (p. 93); bursicon concerned in the hardening and darkening of cuticle (p. 12); the diuretic hormone of *Rhodnius* and other insects (p. 72); a hormone controlling reabsorption from the rectum (p. 72); the 'metabolic hormones' such as the hyperglycaemic hormone and the adipokinetic hormone in *Periplaneta*. Neuro-secretory hormones seem mostly to be small peptides.

Neurohaemal Organs

The neurosecretory material may escape from the nerves at different sites. The diuretic hormone of *Rhodnius* is liberated into the blood along the course of the nerves, from bulbous extremities of axons below the nerve sheath. In nerves acting on the heart-beat the hormone is set free at the endings of the axon in the wall of the heart; there is a similar local release of the hormone which increases the plasticity of the cuticle of *Rhodnius* after feeding. But very often the endings of neurosecretory axons are massed together to form a compact organ in which the secretion is liberated from the swollen extremities of the axons. Such organs, which serve as gateways for the escape of neurosecretion, are called 'neurohaemal organs'.

The corpus cardiacum is the classic example of a neurohaemal organ: most of the active hormones which it contains are derived from neurosecretory cells in the brain and are merely stored in the corpus cardiacum. But it also contains some intrinsic neurosecretory cells which supply hormones to the heart muscle.

Along the ventral nerve cord of most insects there are small swellings on the so-called 'ventral nerves': these are neurohaemal organs providing for the escape of secretion from neurosecretory cells in the ganglia, the functions of which are not known.

Endocrine Glands

The true endocrine glands of insects, like those of vertebrates, are formed by budding off from the ectoderm in the region of the mouth parts. The best known examples are the corpus allatum, which secretes the juvenile hormone, and the 'ecdysial glands' (ventral glands, prothoracic or thoracic glands) which are the source of the hormone causing moulting. This hormone is usually considered to be ecdysone itself (p. 103) but it is possible that it is another hormone which stimulates the production of ecdysone by endocrine organs elsewhere.

Pheromones

A great variety of chemical substances produced by dermal glands exert an influence upon other members of the same species. Where such secretions are used for communication within the species they are termed 'pheromones' or 'social hormones', because there is a certain parallel between the action of such secretions within the community and the action of hormones within the body of the individual.

Pheromones commonly act by stimulating the sense organs of other members of the species and serve to communicate 'alarm', to encourage 'aggegation', or to act as 'trail markers'. They are widely used as sexual attractants or excitants by both sexes but especially by the female (p. 118). The same secretion (formic acid in ants, for example) may serve as a venom, as a repellent against enemies, and as a pheromone for communication within the society. The same pheromone, for example the 'queen substance' of the honey bee (p. 90), can be used for many purposes.

The sensory effects of pheromones may lead to the stimulation of the endocrine system of the insect and thus indirectly bring about changes in behaviour or in growth by hormonal action. In some cases the pheromone itself may have a direct hormonal effect on the recipient. It may be that caste differences in termites are controlled solely by two pheromones: the pheromone given off by the reproductive caste may be the juvenile hormone itself; while a juvenile hormone inhibitor or a corpus allatum inhibitor may be produced by the soldiers. Juvenile hormone is concerned also in the development of queens in the honey bee.

BIBLIOGRAPHY

BLUM, M. S. and BRAND, R. M. *Am. Zool.*, **12**, (1972) 553–576 (social insect pheromones)
BUTLER, C. G. *Biol. Rev.*, **42**, (1967), 42–87 (insect pheromones: review)
NOVAK, V. J. A. *Insect Hormones*, Methuen, London, 1966, 478 pp.
WIGGLESWORTH, V. B. *Insect Hormones*, Oliver and Boyd, Edinburgh, 1970, 159 pp.

Index